Aquaculture for the Developing Countries

A Feasibility Study

Aquaculture for the Developing Countries

A Feasibility Study

Frederick W. Bell
E. Ray Canterbery
The Florida State University

Ballinger Publishing Company • Cambridge, Massachusetts
A Subsidiary of J.B. Lippincott Company

This book is printed on recycled paper.

Library of Congress Cataloging in Publication Data

Bell, Frederick W.
  Aquaculture for the developing countries.

  Includes bibliographies.
  1. Underdeveloped areas—Aquaculture. 2. Technology transfer. I. Can-
terbury, E. Ray, joint author. II. Title.
SH135.B44   338.3'71'091724   76-28400
ISBN 0-88410-296-3

International Standard Book Number: 0-88410-296-3

Library of Congress Catalog Card Number: 76-28400

Printed in the United States of America

Contents

List of Figures

List of Tables

Preface

While the United States and other affluent nations have a more than adequate food supply, other nations daily face the spectre of starvation. The average American or European cannot easily relate to the reports of food shortages and thousands dying from a lack of food. It is difficult to believe. And, unless faced directly with the problem, we are likely to repress it as a normal psychological response. To say that the world is presently facing a critical population—food dilemma of potentially major proportions is so repetitious that most of us have become insensitized to the problem. It exists somewhere else. Or, it's someone else's problem. Direct and indirect efforts to slow population growth—if sufficiently strong—can mitigate the food shortage problem. It is to be hoped that this will occur. If not, the Club of Rome's prediction of a general collapse in world per capita food production may take place early in the Twenty-First Century.

Presently, production from the sea and land is not keeping pace with world population which is doubling every 35 years. The continuation of the Industrial Revolution and scientific advances still finds over two-thirds of the world population facing the grim prediction of Thomas Malthus which led to economics being called the "dismal science." Since people die in the short run or more precisely, today and tomorrow, of starvation, is there anything we can do now? Each margin of effort may count.

This book attempts to assess the usefulness of aquaculture or the farming of aquatic animals and plants in helping to increase world protein. The research is centered on the thesis that through the transfer of rather simple aquaculture technologies from one part of the world to another, starvations and malnutrition may be reduced now. This book attempts to discern areas of the world that suffer from food shortfall, but at the same time have high potentials for adopting aquaculture because of climate, availability of land or coastal shore and relatively low prices of inputs. To the authors' knowledge, this is the first attempt to take a comprehensive look at the potential of aquaculture for the developing countries. We have directed our research toward the small farmer in developing areas that have found that the prescriptions of developmental economics have not proven to work in the short run; that being, to reduce the farm population through rapid growth in agricultural productivity and to industrialize the rest of the developing economy. This prescription has led to massive pools of unemployed and starved people in many urban areas throughout the developing world.

This book is written to serve a varied audience. The governments of developing countries should find it a useful guide in formulating strategies to raise per capita food production and increase foreign exchange earnings. The United Nations should find it helpful in developing assistance programs. It is intended also to help those agencies of affluent governments planning foreign aid, such as the United States Agency for International Development. The strength of technology transfer is that it does not call for massive capital investment or elaborate aid programs, but merely a dissemination of knowledge to small farmers in the developing world. Obviously, we do not expect those farmers to read this book; however, we do hope that extension agents or other practitioners in developing areas will translate our recommendations into realistic plans for most of the world that is food-short. There must be a new awakening of the affluent nations to this problem that will ultimately involve them in a world of increasing interdependence. We offer our assessment of aquaculture as one step in reducing the dimensions of the food problem.

Frederick W. Bell
E. Ray Canterbery

Acknowledgments

The authors would like to thank the intellectual father of this study, Mr. Harold Goodwin, former Deputy Director of the United States Sea Grant Program for his help in stimulating this study and Lee Fletcher and Dan Bromley, then of USAID, for realizing the need for such an assessment. During the course of our aquaculture investigation, members of the research team traveled extensively in order to gain data and insight from internationally-known experts on international aquaculture operations. For example, Dr. Krishna Kumar visited the USAID mission in New Delhi and conferred with Dr. B. Sen and with Professor P.C. George (Joint Commission of Fisheries, Ministry of Food and Agriculture). Professor George was extremely helpful in briefing Dr. Kumar on many Indian aquaculture operations. At the suggestion of Professor George, Dr. Kumar met with Dr. V.G. Jhingran, Director of the Central Inland Fisheries Research Institute. Dr. Kumar also obtained cost and earnings data from Mr. M. Ranadhir and Mrs. T. Rajyalakshmi of the Central Inland Fisheries Research Institute in India. Drs. Charles Rockwood (FSU) and Kumar met, through Dr. John Chang, Mr. Harold Harris of USAID in South Vietnam. Mr. Le Van Dang, Director of Fisheries of the Vietnam Government, arranged for a visit to three fish farms involving the culture of carp.

Of substantial significance, Dr. Kumar met with Mr. Tapia Dar, Director of IPFC in Bangkok, who furnished valuable cost and earnings data for aquaculture operations throughout Southeast Asia.

Dr. Rockwood conferred with Dr. John E. Bardach at the Coconut Island Marine Research Station in Hawaii. Dr. Bardach suggested many approaches to and ways of obtaining information for the study. Dr. Yung C. Shang of the University of Hawaii also helped Dr. Rockwood on' the economics of aquaculture. Later, in Dr. Rockwood's trip to Southeast Asia, he met with Dr. T.P. Chen and others of the Joint Commission for Rural Reconstruction of Taiwan regarding aquaculture operations. The authors would like to express their gratitude to both Dr. Kumar and Dr. Rockwood for the data and the contacts made in the completion of this study.

The authors visited Venezuela, Columbia and Mexico to gain knowledge on the extent to which aquaculture could be seriously considered as a major protein producing sector. The meetings with American officials in these countries confirmed the fact that Central and South America were not moving toward fish culture despite their food population dilemma. Dr. Randy Martin, our research assistant, visited General Mills macrobracium farms and local catfish farms in Honduras. He came away with the conviction that Central America has environmental conditions conducive to many aquacultured species. In this travel phase of the research effort, the researchers visited the following countries: Columbia; Honduras; India; Mexico; Philippines; South Vietnam; Taiwan; Thailand; and Venezuela.

The authors would like to acknowledge especially the help Dr. Randy Martin who served as coordinator for biology and the environment. Also, Dr. Edward Fernald (Director) and Mr. Robert Glassen (Research Associate) of the Florida Resources and Environmental Analysis Center of Florida State University were of great help in the identification of environmental conditions for the various developing countries considered for aquaculture transfer. Professor Phillip Sorensen of the Economics Department contributed to the study estimates of the cost of producing blue-green algae. Mr. Gary Enoch performed the analysis on consumer acceptability of fish in developing countries. Finally, we would like to thank Messrs. Manley Johnson, Bernard Schmitt, William Devouge, Steve Payne, Robert Palmer and other research assistants for their contribution. Of course, our secretaries, Ms. Jacqueline Barocas and Ms. Kerri Wirs, were our most valued employees, as any seasoned researcher knows.

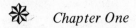 *Chapter One*

The World Food Problem
and Technology Transfer

THE PROBLEM: THE FOOD CRISIS

The pressure of world population on natural resources has
once again given rise to the spectre of Malthusian stagna-
tion and even the possibility of the collapse of civilized life
on our planet. The ghost of Malthus will not be laid to rest. Each
year 10 to 30 million people are expected to die of starvation or of
diseases made fatal by malnutrition. The West African nations, India,
Pakistan, Indonesia and the Philippines are devastated, alternately by
torrential floods and by blistering droughts that produce crop
failures in an unpredictable pattern. Sub-Saharan Africa—often
caught in the grip of long droughts—suffers a death toll that can
climb to several million in a matter of months. The terrible tragedy
underlying today's food/population dilemma is that even if we
succeed in feeding the famished we may only be deferring the
starvation of far greater numbers *unless* food supplies keep pace with
or exceed population growth. In the wake of mounting difficulties
with producing enough food from land areas throughout the world,
more and more attention is being given to alternative sources of
food. It is within this context—the world food crisis—that this study
explores the role of *aquaculture*, or the farming of aquatic animals
and plants, in augmenting the world's food supply. Let us first look
at some of the factors that have been involved in the food/popula-
tion dilemma.

1

WORLD POPULATION GROWTH

The human population is presently increasing at approximately 2 percent a year. Figure 1-1 shows the growth of human population over the last 2000 years compared to bacterial growth (i.e., one cell organisms supplied with enough food, space and water). While bacteria are subject to limits such as space, feed or viruses, humans have been able to overcome many limits to growth. The death rate per 1000 was about 50 nearly 2000 years ago. Today it stands around 10 per 1000 of population. Improvements in medical science and dietary habits have decreased the death rate and have been principally responsible for the acceleration in population growth. World population is now doubling every thirty-five years. Of even greater significance, the areas of the world where food supplies are scarce are precisely those areas where population growth rates exceed the world average, especially Central and South America and North Central Africa. Table 1-1 shows the population growth rates for selected developing countries.[1] At present growth rates the populations of Columbia, Ecuador, Venezuela, Paraguay, Algeria, Pakistan and the Philippines will double in about two decades.

As we find it, the world is divided into two rather distinct groups: one rich, one impoverished; one literate, one mainly illiterate; one well-fed, one malnourished. Population growth, in itself, is not the sole cause of poverty. A few developing countries have managed to sustain a high growth rate of gross national product which has outstripped a high rate of population growth. However, these countries are exceptions to the rule.

WORLD FOOD SUPPLIES

There are more controversies, even among experts, about the potential for food production than about any other factor in this crisis. In the United States, for example, agricultural productivity has been among the highest for any economic sector. Only Canada, Australia, New Zealand, Argentina and Uruguay can match the U.S. standard of diet. And yet, these countries constitute only 9 percent of the world population. Most experts agree that to avoid mass starvation, North America, as the only major food surplus area, must contribute heavily to the nations where food is scarce for a considerable period.

---

1. For the reader's convenience, the nations are divided into two lists that have sampling relevance for our study (see Table 1-2).

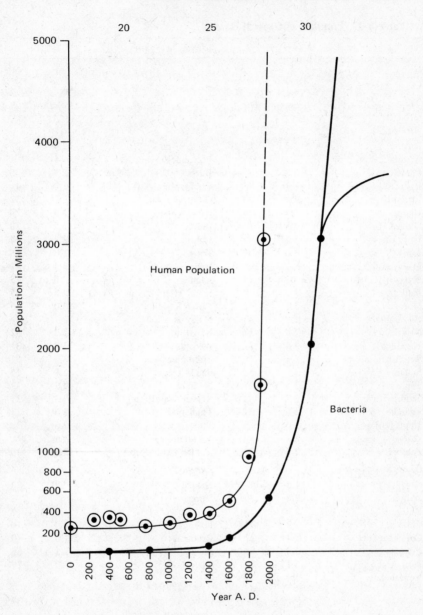

Figure 1-1. World Population Growth Compared to Bacterial Growth (see Frazer, 1971)

Table 1-1.   Population Growth Rates

List 1: Developing Countries (69)					
	% Annual Pop. Growth	Years to Double Pop.		% Annual Pop. Growth	Years to Double Pop.
Africa (24)			*Asia* (25)		
Algeria	3.3	21	Afghanistan	2.4	29
Angola	2.1	33	Bangladesh	NA	NA
Cameroon	2.0	35	Burma	2.3	30
Egypt	2.1	33	China (Mainland)	1.7	41
Ethiopia	2.1	33	China (Taiwan)	2.2	32
Ghana	2.9	24	Hong Kong	2.4	29
Guinea	2.3	30	India	2.5	28
Ivory Coast	2.4	29	Indonesia	2.9	24
Kenya	3.0	23	Iran	2.8	25
Madagascar	NA	NA	Iraq	3.4	21
Malawi	2.5	28	Khmer	3.0	23
Mali	2.3	30	Korea (Dem. Peoples Rep.)	2.8	25
Morocco	3.4	21	Korea Republic	2.0	35
Mozambique	2.1	33	Malaysia	2.7	26
Nigeria	2.6	27	Nepal	2.2	32
Rhodesia	3.4	21	Pakistan	3.3	21
South Africa	2.4	29	Philippines	3.3	21
Sudan	3.1	23	Saudi Arabia	2.8	25
Tanzania	2.6	27	Sri Lanka	2.2	32
Tunisia	2.2	32	Syrian Arab Republic	3.3	21
Uganda	2.6	27	Thailand	3.3	21
Upper Volta	2.0	35	Turkey	2.5	28
Zaire	2.1	33	Vietnam	NA	NA
Zambia	2.9	24	Yemen (Arab Rep.)	2.8	25
South America (8)			*Europe* (7)		
Argentina	1.5	47	Greece	.8	87
Bolivia	2.4	29	Hungary	.3	231
Brazil	2.8	25	Poland	.9	77
Chile	1.7	41	Portugal	1.0	70
Columbia	3.4	21	Romania	1.0	70
Ecuador	3.4	21	Spain	1.1	63
Peru	3.1	23	Yugoslavia	.9	77
Venezuela	3.4	21	*N.C. America* (5)		
			Cuba	1.9	37
			Dominican Republic	3.4	21
			Guatemala	2.6	27
			Haiti	2.4	29
			Mexico	3.3	21
World	2.0	35			

Table 1-1. *(cont.)*

List 2: Developing Countries (21)					
	% Annual Pop. Growth	Years to Double Pop.		% Annual Pop. Growth	Years to Double Pop.
Africa (7)			*Asia* (5)		
Chad	2.3	30	Cyprus	.9	77
Congo	2.1	33	Israel	2.4	29
Dahomey	2.6	27	Jordan	3.3	21
Gabon	.8	87	Laos	2.5	28
Libyan Arab Republic	3.1	23	Lebanon	NA	NA
Niger	2.9	24			
Sierra Leone	2.3	30	*Europe* (2)		
			Albania	2.8	25
N.C. America (5)			Iceland	1.2	58
Costa Rica	2.7	26			
El Salvador	3.2	22			
Honduras	3.2	22	*South America* (2)		
Nicaragua	2.9	24	Paraguay	3.4	21
Panama	2.8	25	Uruguay	1.4	50

Source: Population Reference Bureau, Inc. (1973).

Presently 80 to 93 percent of the world's acceptable agricultural land is being used. Worldwide expansion of agricultural hectarage was a principal source of increased production until 1950, but the production increment from this source fell to only one-fifth during the past decade. Walter R. Schmitt in *The Planetary Food Potential* (1970) points out that the conversion of one-half of the range land and the land used for commercial crops to food production would give a 165 percent increase in world food crop land. But these improvements, obviously requiring enormous capital and causing great dislocation in the world economy and in dietary habits, would help us only for about sixty years at present population growth rates.

Figures 1-2 and 1-3 illustrate the inability of developing countries to raise their per capita production of food relative to the developed countries over the 1961-73 period. While more developed countries have experienced substantial increases in both *aggregate* and *per capita* food production, developing countries have matched the developed areas in increasing aggregate food production, but have failed to increase *per capita* food production, which remains at a subsistance or less than subsistance level. In countries containing one-third or more of the world's people, average food intake today is below the minimum required for normal growth and activity. Much

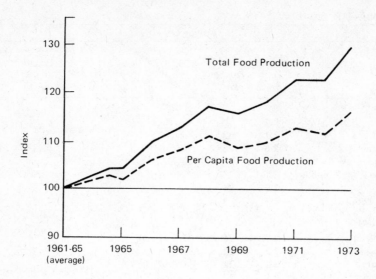

Source: U.S. Dept. of Agriculture

Figure 1-2. More Developed Countries: Total and Per Capita Food Production Rises Substantially

of the population in these countries suffers from chronic protein malnutrition.

According to Meadows *et al.* (1972), the present rate of resource exploitation, coupled with exponential population growth, will result in a collapse of the world's economic system in terms of food production per capita. They argue that despite technological advances, resources are limited and the current population explosion will produce this collapse regardless of attempts to improve food supplies. Hence, there is a recognition that food production improvements must be coupled with a substantial reduction in the rate of population growth. In this book we address ourselves to the food supply side of the problem.

FAMINE AND CHRONIC FOOD SHORTAGES

Presently there is a belt of hunger around the equatorial bulge. Nearly all the persons in this region are ill-fed, at least one-half are malnourished and perhaps as many as one-fifth are starving. These nations are listed separately for they represent a special need for augmented protein production.

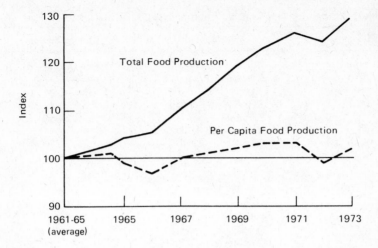

*Excluding Communist Asia

Source: U.S. Dept. of Agriculture

Figure 1-3. Developing Countries*: Population Growth Absorbs Food Production Increases

Table 1-2 is divided among those nations that were suffering famine in 1973, those near famine, those experiencing potential famine and those that chronically faced food shortages. Those experiencing famine in 1973 were the drought-stricken nations of north and central Africa. And, one-half of those "near famine" were African nations. Those in the "potential famine" category are only one bad harvest away from starvation conditions while those with chronic food shortages are always struggling with agricultural production in order to avoid potential famine.

Of the 33 nations on this list, only six are excluded from the sample (discussed below) used in our simulation study. And these excluded nations are contiguous with nations in the sample so that it would be possible to draw some inferences regarding them from "neighborhood" results. Most of these nations have long been ranked low in terms of per capita protein production. Nonetheless, until recently the prospects for increased protein per capita from grain had appeared favorable. That prospect has been substantially altered—at least temporarily.

The Green Revolution, which had brought India, Pakistan and other Asian nations to self-sufficiency in grain production (Canter-

Table 1-2.   Famine and Chronic Food Shortages: 33 Nations

Famine	Near Famine	Potential Famine
Chad	Bolivia	Cameroon
Ethiopia	Kenya	Guinea
*Gambia	Nigeria	*Guyana
Mali	Syrian Arab Republic	India
*Mauritania	Tanzania	*Yemen, Democratic
Niger	Yemen Arab Republic	
*Senegal		
Upper Volta		

Chronic Food Shortage

Algeria
Angola
Bangladesh
*Central African Republic
El Salvador
Ecuador
Indonesia
Iran
Iraq
Haiti
Philippines
Saudi Arabia
Uganda
Zaire

*Nations not included in study sample.

bery and Bickel, 1971) has suffered a setback that has contributed to the food problem. Increases in world oil prices have nearly tripled the cost of nitrogen fertilizers and fuel for irrigation pumps, two inputs that are absolutely essential for continued success with the high-yielding grain varieties. Small farmers are now reverting to the old varieties that can be grown with traditional methods. These reactions to rising input prices go a long way toward explaining the present conditions in some of those nations now experiencing potential famine or chronic food shortages.

FOOD FROM THE SEA

Production from the sea has always been offered as a panacea for increasing food production. Estimates (FAO, 1969) indicate a maximum potential of 120 million metric tons of fish from *wild stocks.* Presently, we are annually harvesting around 65 million metric tons, and projections (Bell, *et al.*, 1971) indicate that by the year 2000 fish consumption will increase by around 50 percent (i.e., from wild stock species only) using very conservative assumptions about population and income growth. This consumption compares with a projected 67 percent growth in population. Therefore, the wild fishery stocks of the sea do not offer anything near a panacea, but only a mitigating force in supplying increased protein to the world.

The pinch between declining available agricultural land and the limit on wild fish stocks has caused some food experts to consider aquaculture as a possible source of protein. *Aquaculture is defined for the purposes of this book as the culture and husbandry of aquatic organisms; the control and management of aquatic plants and animals reared in large numbers in controlled or selected environments for economic or social benefits.*

The cultivation of fish, mullusks and crustaceans under controlled conditions is a promising source of high quality food. According to a report by Ryther and Bardach (1968):

> a ton or more of fish and a hundred tons of shellfish can be raised in the same space that it takes to raise a few hundred pounds of beef cattle. Intensive pig farming in developed nations leads to a production per man-year of around twenty-five tons of live pigs while an oyster farmer can raise forty to sixty tons of oysters (shells excluded) per year. The average Danish trout farm produces about forty tons of trout per year, with two or three men employed to do the labor. The sewage ponds of the Bavarian Power Company, near Munich, have the capability of producing 100 tons of carp from about 200 hectares (about 500 acres) of water. Three men tend the ponds and the fish; thus the fish production per man-year would exceed 30 tons.

According to some studies, commercial culture of oysters could yield 170 pounds per acre per year average, compared to the six pounds obtained from natural public beds, and suggests that yields could go as high as 5000 pounds under optimum conditions in the United States or to the 50,000 pound maximum of Japan. Shrimp, in Japan, are raised in yields of 9,000 pounds per acre, under forced culture, in a period of ten months. Since some fish feed on unused products and even waste, there is a possibility of tremendous yields at low costs. These yields moreover are from a technology currently in its infancy with very little research on genetics and the optimization of conditions. Before enthusiasts run wild with extrapolation, however, it must be pointed out that most aquaculture can be done only where conditions are favorable, where shallow, unpolluted estuaries or lagoons are available at reasonable cost, or, with fresh water species, where large shallow ponds are feasible and ample water is at hand. Within the context of the population/food problem, we will consider the potential of aquaculture to augment the world's food supply.

THE RESEARCH PLAN

The main objective of this study is to test the economic feasibility of transferring existing aquaculture technologies from one area of the world to those developing countries that are short of food. We assume that the main goals of such transfers are:

1. Increased domestic food production and consumption;
2. Increased domestic employment, and;
3. Expanded exports.

One main emphasis of this study will be the small farmer. That is, while the successful farmers of the world are producing in the wake of what has been called the "Agricultural Revolution," more than half of the world's farmers are still farming in the ways of their ancestors—barely eking out a subsistence. In the developing areas of the world people have no alternatives: there are few industrial jobs for them to escape to, nor would they be qualified if there were. Due to rapidly growing population in developing countries and migration to the cities, starvation, poverty and degradation exist. Some way must be found to make it possible for small farmers to stay on the land and make a decent living there. The share of the population in farming among developing countries is slowly going down, but the absolute numbers are going up, despite migration. Eventually technological progress is bound to push many of these people out of farming. But until it does and until there are other jobs to go to—the question simply becomes: do we try to help these people or not?

Robert F. McNamara, President of the World Bank, has pointed out that "if the poor farmers in developing countries do not benefit from increased food production (via the Green Revolution for example) the whole drive toward greater production and employment will be diminished by a sluggish market."

In assessing the potential for aquaculture to achieve these *three* main objectives, two types of assessments will be conducted.

1. An evaluation of aquaculture protein sources compared with alternative sources that will yield data and judgments on:
    (a) the type of product and quality of protein;

    (b) the production possibilities, available resources, and possible volume;

    (c) consumer acceptance and demand in relation to anticipated prices; and

    (d) the contribution to rural employment and income

2. An appraisal of research and development requirements for aquaculture enterprises in developing countries:

    (a) labor requirements;

    (b) research and development opportunities; and

    (c) the direct and indirect effects on rural development and nutrition

Steps in the Research Plan

Step 1—In order to implement this study as well as to further the research in this field, an annotated bibliography of the aquaculture literature was prepared. We consulted with the National Marine Fisheries Service for their accomplishments and knowledge as well as the Office of Sea Grant in preparing the bibliography.

Step 2—Through the use of the bibliography and other information we established a *list* of the principle species or plants presently being aquacultured in the world today (see Chapter 2). The following several criteria were used in selecting from this list, species or plants for technology transfer:

    (a) Contribution to protein production or export in an existing enterprise in a particular country (i.e., is the aquaculture operation fairly viable?);

    (b) Environmental conditions (temperature, salinity, PH factor, etc.)

    (c) Potential consumer acceptability;

    (d) Adaptability to small farm operations;

    (e) Contribution to employment in an existing enterprise in a particular country.

The purpose is to compile a list of animal species or plants (primarily from developing countries, but not necessarily) that will have the highest potential for and benefits from technology transfer from one area of the world to various developing countries which presently do not culture the product. According to T.V.R. Pillay (1973) of the United Nations Food and Agriculture Organization, world aquaculture production is on the order of five million metric tons annually.

The species most heavily cultured are fresh water animals over which life-cycle control is good or marine and brackish species which have a good supply of easily collected juveniles. *Carp*—which accounts for the largest volume among the cultured species—has a limited market in the U.S. but may be acceptable for developing countries. *Milkfish, mullet, oysters, mussels, shrimp, trout, catfish* are among some of the species that are presently cultured.

A successful aquaculture enterprise operation in one developing country does not *insure* a successful transfer. Nonetheless, since developing countries do have many similar characteristics (climate, labor costs, etc.), *a priori* transferability from one developing nation to another appears a reasonable initial hypothesis. Therefore, in addition to a species list, Step 2 also produced a large sample of developing nations to be considered as technology receivers (see below).

Step 3—A generalized bioeconomic model was developed for simulation purposes. (The model is detailed in Chapter 5.) This was the only feasible methodology available to the authors. First, the parameters of each aquaculture enterprise were estimated (see Chapter 3). Second, all variables that go into the decision-making process for deciding the economic feasibility of the particular aquaculture enterprise were estimated for each sample nation (see Chapter 4). Thus, for each species in the sample compiled in Step 2, we modeled an aquaculture enterprise for each sample nation from available data.

Step 4—Secondary, and when necessary, primary input data were gathered on each sample country as it related to the aquaculture enterprises simulated. We merely changed input data (factor prices and environmental shifters) for the aquaculture enterprise to conform to the potential developing nation receiver. (The input data are provided in Chapters 4.) A computer simulation generated the following data:

(a) Average cost per unit of protein (for ith species);
(b) Protein yield per unit of land (for ith species);
(c) Employment per unit of protein output (for ith species);
(d) Output per unit of land, and;
(e) Foreign exchange earnings per unit of land.

These results are presented in Chapters 6 and 7.

Step 5—This phase analyzed alternatives to aquaculture presently being used in the sample of developing countries with respect to the following: (1) increased domestic food production and protein consumption; (2) increased domestic employment (mostly rural); and (3) expanded exports as compared to aquaculture.

Step 6—We used the information developed in Steps 1-5 to produce a "recommended receivers" list of where aquaculture might accomplish the stated objectives (see Chapters 6-8).

Let us now consider the sample of developing countries selected for consideration in this study.

SELECTION OF DEVELOPING COUNTRIES FOR AQUACULTURE TECHNOLOGY TRANSFER

A minimum sample of developing nations was compiled using strictly objective criteria. The countries in this sample (Sample 1) meet two criteria: (1) per capita Gross Domestic Project at 1970 constant market prices is less than or equal to one-fourth that of the United States in 1970 ($1200) and; (2) population is equal to or greater than four million in 1970. Sample 1 is comprised of 69 countries.

The main reason for eliminating various countries from Sample 1 are: (1) Nations with populations below four million have relatively small domestic markets; (2) Such nations comprise a small share of the total population of developing nations; and (3) Economies of scale on public-assistance funding probably will exclude most very small nations.

A second sample was computed of developing countries that failed to qualify under the population criterion of Sample 1, but nonetheless appear on most developing nation listings (such as those of Adelman-Morris or the U.N.) as being important nations that may have on-going aquaculture programs or the potential for such programs. Sample 2 is comprised of 21 countries. The addition of Sample 2 brings the total number of developing countries under study to 90 which is slightly larger than the "textbook" lists of 74-75 and the USAID 1961 list of 84. Small islands have been excluded from both lists, because of the smallness of their populations.

A third sample, Sample 3, is a select group of 12 developed nations that produce substantial quantities of food. These nations are studied for clues to their successes in protein production that may be useful in considerations of technological transfers to developing nations.

Those country samples defined from these criteria appear in Table 1-3. Note that the number of nations in each region on each list appears in parentheses. Within each regional grouping, nations are ordered alphabetically.

Table 1-3.   Four Sample Lists Used in the Study

Sample 1: Developing Countries (69)

Africa (24)	*Asia* (25)	*South America* (8)
Algeria	Afghanistan	Argentina
Angola	Bangladesh	Bolivia
Cameroon	Burma	Brazil
Egypt	China (Mainland)	Chile
Ethiopia	China (Taiwan)	Columbia
Ghana	Hong Kong	Ecuador
Guinea	India	Peru
Ivory Coast	Indonesia	Venezuela
Kenya	Iran	
Madagascar	Iraq	*Europe* (7)
Malawi	Khmer	Greece
Mali	Korea (Dem. Peoples Rep.)	Hungary
Morocco	Korea Republic	Poland
Mozambique	Malaysia	Portugal
Nigeria	Nepal	Romania
Rhodesia	Pakistan	Spain
South Africa	Philippines	Yugoslavia
Sudan	Saudi Arabia	
Tanzania	Sri Lanka	*N.C. America* (5)
Tunisia	Syrian Arab Republic	Cuba
Uganda	Thailand	Dominican Republic
Upper Volta	Turkey	Guatemala
Zaire	*Viet Nam (Dem. Rep.)	Haiti
Zambia	*Viet Nam (Rep. of)	Mexico
	Yemen (Arab Rep.)	

Sample 2: Developing Countries (21)

Africa (7)	*N.C. America* (5)	*Asia* (5)	*Europe* (2)
Chad	Costa Rica	Cyprus	Albania
Congo	El Salvador	Israel	Iceland
Dahomey	Honduras	Jordan	
Gabon	Nicaragua	Laos	*South America* (2)
Libyan Arab Republic	Panama	Lebanon	Paraguay
Niger			Uruguay
Sierra Leone			

Sample 3: Developed Nations (12)

Australia	Italy
Canada	Japan
Czechoslovakia	New Zealand
Denmark	Norway
France	U.K.
Germany, Fed. Rep.	U.S.A.

Table 1-3. *(cont.)*

Sample 4: Rest of World (1970)

Africa (26)

Botswana
British India Ocean Territory
Burundi
Cape Verde Islands
Central African Republic
Comoro Islands
Equatorial Guinea
French Territory of the Afars and the Issas
Gambia
Lesotho
Liberia
Mauritania
Mauritius
Namibia
Portuguese Guinea.
Reunion
Rwana
St. Helena
Sao Tome and Principe
Senegal
Seychelles
Somalia
Spanish North Africa
Spanish Sahara
Swaziland
Togo

North Central America (24)

Antigua
Bahamas
Barbados
Bermuda
British Honduras

Europe (21)

Andorra
Austria
Belgium
Bulgaria
Faeroe Islands
Finland
German Democratic Republic
Gibraltar

North Central America (Cont.)

Cayman Islands
Dominica
Greenland
Grendad
Guadeloupe
Jamaica
Martinique
Montserrat
Netherlands Antilles
Panama Canal Zone
Puerto Rico
St. Kitts-Nevis-Anguilla
St. Lucia
St. Pierre and Miquelon
St. Vincent
Trinidad and Tobago
Turks and Caicos Islands
Virgin Islands (U.K.)
Virgin Islands (U.S.)

Asia (14)

Bhutan
Brunei
Gaza Strip
Kuwait
Macau
Maldives
Mongolia
Oman
Portuguese Timor
Qatar
Sikkim
Singapore
United Arab
Yemen, Democratic

Oceania (25)

American Samoa
British Solomons
Canton and Enderbury Islands
Christmas Island
Cocos Islands
Cook Islands
Fiji

Table 1-3. *(cont.)*

Sample 4: Rest of the World (1970) (continued)

Holy See	French Polynesia
Ireland	Gilbert Islands
Liechtenstein	Guam
Luxenbourg	Johnston Island
Malta	Midway Islands
Monaco	Nauru
Netherlands	New Caledonia
San Marino	New Hebrides
Sweden	Niue Islands
Switzerland	Norfold Island
UK Channel I	Pacific Islands
UK Isle Man	Papua and New Guinea
USSR	Pitcairn Island
	Tokelau Islands
South America (4)	Tonga
Falkland Islands	Wake Island
French Guinea	Wallis and Tutuna Islands
Guyana	Western Samoa
Surinam	

*Our data for modeling are from 1970 in which two Viet Nam's were reporting.

A fourth list, Sample 4, includes all those nations of the world *excluded* from Lists 1, 2, and 3. This final list provides immediate information on exclusions.

POPULATION ANALYSIS OF SELECTED SAMPLES

The following analysis indicates the coverage of the sample in terms of world population. Our criteria classify some countries as developing which the United Nations classifies as developed (e.g., South Africa). Also, the U.N. constructs economic classes of "developed market economies," "developing market economies," and "centrally planned economies" rather than developed vs. developing designations. As a consequence of these different classifications, our total population in developing countries is greater than the total derived by adding the U.N. category of "developing countries" plus the "planned" countries of China, Democratic Peoples Republic of Korea, Mongolia, Democratic Republic of Vietnam, Albania, Bulgaria and Romania. (China, in the U.N. data, excludes Taiwan.)

Because of the slightly different definition of "developing" and because level of development is broadly associated with continental regions, we have computed the 1970 population shares represented by countries in Samples 1 and 2 by continent. The smaller shares of sample population in North America and Europe reflects development rather than noninclusiveness of our developing nations sample. The populations and shares by region appear in Table 1-4.

The small developing countries excluded from Samples 1 and 2 had a total 1970 population of only about 50 million. If we use the U.N. classification, total developing countries (1970) population was 2,560,000,000. With the U.N. data as base, our samples exclude slightly less than 2 percent of the developing countries' populations.

In terms of world population, Sample 1 includes 72.15 percent of the world total and Samples 1 and 2 combined include 73.45 percent of the total. Because of our slightly divergent definitions these shares exceed the commonly accepted figure that has two-thirds of the world's population living in developing countries.

SUMMARY

Starvation and malnutrition stalks the developing world. The planet is rapidly exhausting its supply of tillable soil. Many experts look seaward for a reprieve from the food crisis. But the wild stock fisheries also is a finite resource. Aquaculture, which is an intensive

Table 1-4. Share of Total Population Contained in Samples 1 and 2 by Continent

Region	1970 Population	Population Covered	Shares of Total Covered
Africa	344,343,000		
Sample 1 (24)	302,314,000	318,509,000	92.5%
Sample 2 (7)	16,276,000		
N. America	319,301,000		
Sample 1 (5)	73,734,000	85,179,000	26.7%
Sample 2 (5)	11,445,000		
South America	190,053,000		
Sample 1 (8)	183,565,000	188,857,000	99.4%
Sample 2 (2)	5,292,000		
Asia	2,040,092,000		
Sample 1 (25)	1,914,938,000	1,926,552,000	94.4%
Sample 2 (5)	11,614,000		
Europe	461,843,000		
Sample 1 (7)	135,550,000	137,886,000	29.9%
Sample 2 (2)	2,336,000		
Sample 1 (total = 69)	2,610,101,000	2,657,064,000	73.5%
Sample 2 (total = 21)	46,963,000		
World	3,617,362,000		

use of water resources, may offer a substantial alternate source of world protein. The potential is sufficient to have motivated this study of the feasibility for aquaculture technology transfer.

We have identified a large sample of developing nations to consider as potential technology receivers. These nations include virtually all the population of the developing world. In the next chapter we will discuss aquaculture's current role and select the species for possible transfer to our country sample.

REFERENCES

Bell, W. *et al.*, "The Future of the World's Fishery Resources to the Year 2000," Preprints, 7th Annual Conference, Marine Tech. Soc., 1971, Washington, D.C.

Canterbery, E.R. and Bickel, H., "The Green Revolution and the World Rice Market, 1967-1975," *American Journal of Agricultural Economics* (May 1971).

FAO, "The Prospects for World Fishery Developments in 1975 and 1985," FAO Indicative World Plan, Rome, 1969.

Frazer, D., *The People Problem*, Bloomington: Indiana University Press, 1971.

Meadows, D.H., Meadows, D.L., Rauders, J. and Behrens III, W., *The Limits to Growth*, New York: Universe Books, 1972.

Pillay, T.V.R., "The Role of Aquaculture in Fishery Development and Management," FAO Conference on Fishery Management and Development. Vancouver, 1973.

Ryther, J.H., Bardach, J.E., *The Status and Potential of Aquaculture*, Clearinghouse for Federal Scientific and Technical Information, P.B. 177, 768 (Springfield, Virginia, 1968).

Schmitt, W.R., "The Planetary Food Potential," *Annals of the New York Academy of Sciences*, Vol. 118, pp. 645-718 (1970).

Statistical Yearbook, Food and Agricultural Organization of the United Nations, 1970, 1971, 1972.

World Population Data Sheet, Population Reference Bureau, Inc., 1973.

✳ *Chapter Two*

The Current Role of Aquaculture: Potential for Species Transfer

INTRODUCTION

Aquaculture is a very old and highly productive management practice. As examples, the culturing of oysters and Chinese carp probably have been conducted since at least Roman times. Aquaculture yields vary considerably with the species cultured and with the type and intensity of the management or farming practice. But aquaculture yields typically are several times the output that would occur naturally.

The purposes of aquaculture are extremely varied, and not always or even primarily provision of an economical source of protein. Even so, aquaculture currently provides an estimated 10 percent of the world's water-derived proteins and about 3 percent of all the world's protein, exclusive of milk. World aquaculture production now is estimated at 5 to 6 million metric tons, with a current value in excess of U.S. $2.5 billion (Caton, *et al.*, 1974).

Several facts portend potential for aquaculture in many regions. Its concentration in the Indo-Pacific region appears to be based more upon culture than environment. The aggregate production trend is favorable: aquaculture output approximately doubled in the five years, 1969-1974 (Caton, *et al.*, 1974). And, some select countries already rely upon aquaculture for up to 50 percent or more of their total fisheries production.

It is clear from the foregoing that world aquaculture is well beyond the purely experimental stage. Presently profitable technologies exist that can be and are being transferred within and among nations.

19

In this chapter we will find the reasons for the recent and projected rapid growth of aquaculture activities to be: (1) the growing and incessant world demand for protein; (2) the rising costs of commercially caught fish, as maximum sustainable yields of more and more wild stock species are approached or exceeded; and (3) the favorable feed conversion rates and high productivity per hectare of aquaculture operations, as compared with traditional agricultural methods and products.

Nonetheless, aquacultural practices seem to be more complex and require greater husbandry skills than other forms of agriculture. Also, the transfer of aquaculture technology within and between nations is not nearly so simple as is the transfer of row cropping or animal husbandry technologies. Optimum feeding programs, for example, often vary quite widely with moderate changes in environmental constraints, local food constraints, local food preferences and the relative cost of the necessary inputs—feed, fertilizer, labor and pond construction, primarily. And, finally, there is the question of species selection for possible technology transfer.

The number of choice sets facing the aquaculturist who must adjust to his own particular physical, economic and cultural environment is thus extremely large and a good deal of research, which would incur costs, would be of a non-patentable nature, even if successful. The chances of such necessary technology transfer efforts occurring in adequate amounts, especially within the impoverished regions of the world, is unlikely in the absence of special governmental efforts to foster it. In particular, entreprenurial activity on a relatively large scale is beyond the unaided financial and technical capacities of the small farmer or rural businessman, who is usually undercapitalized, and almost always lives out his life in a highly competitive environment that neither forgives mistakes nor rewards one financially for general contributions to the stock of knowledge. Nonetheless, this high risk does not necessarily rule out adoption on a small scale with only a minimal capital assistance.

THE EXTENT AND DISTRIBUTION OF WORLD AQUACULTURE

The exact extent of world aquaculture is not known. Partly the problem is definitional. At what point does wise and careful management of natural resources shade off into farming and husbandry of freshwater and marine organisms? But mostly the problem is simply one of measurement. Many aquaculture operations—like those of agriculture, in general—are carried out in remote areas and the product is either consumed on the farm or bartered locally and in any event never included in national or world production figures.

On the basis of data collected from 36 countries by the FAO (1970), world aquaculture output of fin-fish in 1970 was estimated to have been over three million tons. The estimated production of mollusks for that same year was put at one million tons, and that of seaweeds at about one-third of a million tons. Thus total world aquaculture output in 1970 seems to have been in excess of five million tons, exclusive of the production of sport fish and bait fish, ornamental fish, and cultured pearls. The precise extent of sport and bait fish culture apparently is not known, but is obviously a large and growing activity. Ornamental fish culture must also be large, but again is not measured. It is estimated that the U.S. alone imports over $80 million worth of ornamental fish annually. Ornamental fish culture, too, is a large and expanding activity. In 1969 Japan produced over 100 tons of pearl oysters, valued at more than $500 million (Furukawa, 1971).

In a more recent study Pillay (1973) estimated fin-fish aquaculture production in 42 countries, including mainland China, at about 3.7 million metric tons, shrimp and prawn aquaculture production for seven countries at 14 thousand tons, oyster culture in 11 countries at 710 thousand tons, mussel production in five countries at 209 thousand tons, cultured clam output in three countries at 57 thousand tons, and miscellaneous mollusk production in one country at 20 thousand tons, while seaweed culture in two countries was estimated at 373 thousand tons. Total measured aquaculture output for these selected countries and products was thus slightly above five million metric tons, annually (see Tables 2-1 and 2-2).

The total area used in world aquaculture production is also not known, but statistics on fin-fish production indicate that average production per hectare is about 1.5 metric tons per year. Based upon this estimate there are more than 2.6 million hectares of water impoundments used in world aquaculture production. This sum is exclusive of the area used for mollusk and seaweed production, which most often is open ocean or open estuarine area anyway.

Trends in aquaculture production are even more difficult to find than current production statistics. Whether or not the Caton, *et al.* estimate of growth is reasonably precise, a very high rate of growth in aquaculture production could be sustained for a considerable period merely by extending present technology to additional areas and by raising the output of existing areas to currently feasible levels. The Caton *et al.* prediction of potential aquaculture output of at least 50 million metric tons by the year 2000 seems quite feasible in

Table 2-1. World Fish and Shellfish Production

Country	Total Catch	Aquaculture Fin-Fish Production	Aquaculture Fin-Fish % of Total Production	Aquaculture Fin-Fish Area (hectares)
China (mainland)	6,880,000	2,240,000	32.6	700,000
India	1,845,000	480,000	26.0	607,915
U.S.S.R.	7,340,000	190,000	2.6	126,666 (est.)
Indonesia	1,249,700	141,075	11.3	266,300
Philippines	1,049,700	94,573	9.0	164,414
Thailand	1,571,600	87,764	5.6	58,509 (est.)
Japan	9,994,500	85,000	0.9	508
Taiwan	650,200	56,185	8.6	39,234
U.S.A.	2,766,800	40,200	1.5	28,300
Pakistan & Bangladesh	416,500	37,540	9.0	30,780
Malaysia	390,300	25,648	6.6	90,473
Hungary	26,000	19,697	75.8	22,000
Italy	391,200	18,000	4.6	12,000 (est.)
S. Vietnam	587,500	16,500	2.8	2,500
Yugoslavia	49,200	15,840	32.2	9,747
Ceylon (Sri Lanka)	87,700	15,000	17.1	10,000
Rumania	68,700	12,000	17.5	6,400
Denmark	1,400,900	11,000	0.8	7,333 (est.)
Poland	517,700	10,909	2.1	62,791
Czechoslovakia	13,000	10,641	81.9	42,798
Israel	28,200	10,220	36.2	4,904
Brazil	515,400	9,967	1.9	6,644 (est.)
Mexico	402,500	9,026	2.2	12,650
Khmer	171,100	5,000	2.9	3,333 (est.)
Germany, East	323,100	3,669	1.1	2,446 (est.)
Germany, West	507,600	2,627	0.5	11,824
Burma	442,700	1,494	0.3	2,920
Zaire	145,800	1,406	1.0	4,058
Bolivia	1,200	1,400	–	25,502
Austria	4,000	780	19.5	3,000
Hong Kong	114,100	690	0.6	629
Zambia	39,300	689	1.8	459 (est.)
Uganda	137,000	670	0.5	410
Madagascar	48,000	615	1.3	1,280
Norway	3,074,900	600	.02	400 (est.)
Singapore	15,000	554	3.7	890
Nigeria	155,800	127	0.1	85
Kenya	35,000	122	0.3	610
Spain	1,498,700	50	.003	33 (est.)
S. Korea	1,073,700	40	.003	76
Ghana	220,400	30	.01	204
Puerto Rico	57,700	25	.04	135
	69,400,000	3,657,373	5.3 (avg.)	2,643,551

Source: Based upon Pillay (1973) and FAO Fisheries Yearbook (various).

Table 2-2.   Estimated World Production Through Aquaculture Exclusive of Fin-fish Production

Country	Shrimps and Prawns	Oysters	Mussels	Clams	Other Mollusks	Seaweeds
Australia		9,800				
Canada		4,100				
France		34,200	39,800			
India	3,800					
Indonesia	3,328					
Italy			13,700			
Japan	1,800	194,600		10,800		357,000
Korea		45,700	16,800	16,800	20,000	16,000
Malaysia	250		28,600			
Mexico		43,500				
New Zealand		10,700				
Philippines	2,500					
Portugal	120	2,900				
Singapore		1,800				
Spain	2,500		109,700			
Thailand				60		
Taiwan		12,700				
United States		350,500				
	14,298	710,500	209,600	57,660	20,000	373,000

Source: Pillay (1973).

light of the relatively low technology of much current aquaculture as well as lack of extensive application of the technique. Such a production level also seems desirable in light of present and projected deficiencies in world protein production.

AQUACULTURE SPECIES

There are some 20,000 known species of fish, excluding various aquatic plants, shellfish and other marine life. While most of these organisms are not amenable to complete human control from spawning to market (given current technology), most are amenable to manipulation of at least one stage of life before harvest, for the purpose of increasing production and yield. Many of these species are commercially valuable aquaculture species. For example, milkfish, shrimp and a number of other valuable aquaculture species cannot be spawned in captivity.

From various sources a list of over 200 already cultured species of fin-fish and shellfish was compiled, and it is an incomplete listing. Ling (1972) lists 20 species of fin-fish, 25 species of crustaceans, 20 mollusks and 10 algae already being cultivated in the coastal areas of the Indo-Pacific region alone, with roughly one-third of these being cultured extensively. To the list Ling compiled, one would have to add freshwater species, which were not covered in his study, as well as numerous species cultured outside the Indo-Pacific region exclusively, and hence again not covered by Ling. It is clear that an extremely wide variety of fishes and other marine organisms might be enhanced through aquaculture.

Presently the most extensively cultured species of fish and shellfish, ranked in general order of their economic importance, appear to be as shown in Table 2-3 below. However, this is not necessarily the optimum grouping for the study of aquaculture's potential role in helping to meet world protein needs. Many aquacultured species are not even produced for food. Sport fish and bait fish are not produced primarily for direct consumption as food, many seaweeds are produced only for the carrageenin[1] that can be extracted from them, etc. Also, most aquacultured species produced are for the luxury trade. Species emphasized in current aquaculture are often selected because they command a favorable market price, and not because their cost of production per unit of protein is low.

---

1. Carrageenin is used as a stabilizing agent in various food products, in sizing cloth, leather tanning and in paper making.

Table 2-3. Fin-Fish and Shellfish Species Most Extensively Cultured

Species	Estimated Production
Common Carp ⎫ Chinese Carp ⎬ Indian Carp ⎭	210,000 metric tons in 1965
Tilapia	NA
Shrimp and Prawns	14,298 metric tons
Oysters	710,000 metric tons
Milkfish	167,000 metric tons
Mullet	NA
Clams and Mussels	236,260 metric tons
Clarias	NA

Source: Pillay, T.V.R. (1972, 1973) Bardach (1972).

POTENTIAL ROLE OF AQUACULTURE AS A SOURCE OF WORLD PROTEIN

Fish are a superior source of animal protein. If the favorable amino acid pattern is considered, human beings can utilize at least 83 percent of the raw weight of fish, and animal protein is especially valuable to humans because it contains two essential amino acids (lysine and methionine) not found in adequate amounts in vegetable protein.

A pound of fish can also be produced more cheaply than a pound of red meat. Fish, being cold blooded, take on the temperature of their environment, instead of expending calories to maintain a constant body temperature. Also, fish live in an environment that supports them, while land based animals must use a good deal of energy to develop a skeletal system and in general to support themselves against gravity. For these reasons fish are better feed converters than land based animals. Feed conversion rates for fed fish are about one and one-half times as great as swine or chickens and about twice as great as cattle or sheep (Bardach, 1972). Also fish can be crowded more closely than land based animals because they can utilize the water column in a three-dimensional habitat. Thus, in well-managed environments 2000-3000 kg. or more of fish can be produced per hectare per year while the maximum figure for cattle is 500 to 700 kg. (Delaney and Schmittou). And, while some fish must be fed a ration as expensive as cattle or poultry feedgrains, many varieties of fish and other marine organisms can live very well on nutrients found naturally in their aquatic environment, or fostered through fertilization of their environment, mostly by the addition of phosphates.

Aquaculture often does require more labor and sometimes more capital per unit of food output than other types of farming. But in many cases the labor required for aquaculture operations can be scheduled around the peak labor demand periods for agriculture, as for example at planting and harvesting time, so that fish rearing, particularly on a small scale, can be at least partly complementary with general farming. The land requirements for aquaculture often permit the use of low economic value land—ravines, swampland, saltwater marsh or mangrove area—that is not well suited to other uses. And, the desirable complementarity of water conservation, provision of recreational area, or even flood control helps to further justify the often moderately expensive capital construction requirements.

The natural advantages of fish as a protein source strongly suggest the value of increased use of that product in diet deficient areas of the world. Wild stocks of fish already have been heavily exploited in most areas of the world and are more and more likely to be heavily exploited in the coming years. This suggests the need to turn to some alternative source of fish rather than increasing fishing pressures on finite natural stocks. Given the rising world demand for protein, the National Marine Fisheries Service has estimated that maximum sustainable yields of salmon, halibut, groundfish, crabs, fish meal and lobsters will be reached or exceeded by 1985, and most of the remaining species of commercial importance (except sardines, scallops and clams) will reach this peril by the year 2000. Aquaculture seems a logical alternative source for fish protein production, one to which the world increasingly is compelled to turn.

Present day aquaculture, however, often is not a cheap source of protein, because—as suggested above—much of it is devoted to the production of luxury foods. Fish farming presently is a business, often more profitable than agriculture (Shang, 1973). Not surprisingly, profit-motivated fish farming has concentrated much effort on the production of high value output such as shrimp, prawns, eels, channel catfish, salmon and yellow-tail. These aquaculture products are *not* the cheapest source of animal protein. Their production normally requires feeding of prepared foods, which together with the typically higher labor and capital requirements of aquaculture operations, pegs the average cost of the final product at a high level.

To be an economical source of protein, aquaculture must concentrate on production of species that are near the bottom of the food chain. Feeding fish to fish or even grain to fish is apt to be a losing proposition unless the produced fish has a high market value. If cheap animal protein is the goal, only where trash fish are available in large numbers would one normally want to violate the general rule and feed fish protein to fish. Trash fish as a source of fish feed may be produced in substantial quantities by the richer nations and by those nations whose people do not have a particular fondness for fish. But trash fish are not produced in significant quantities by food-short nations whose people are fish-loving. The Philippinos, to take one example, eat nearly every species of fish, even the very smallest fish, and nearly every part of every fish, including fins, heads and most of the entrails. The Philippine people produce a lot of fish food but not much fish waste.

SPECIES SELECTION FOR THE SIMULATION STUDY

The selection of species or aquaculture farm for transfer from one area of the world to another at first glance appears extremely complex. An obvious problem for the researcher is in the nature of pre-simulation selection of an aquaculture farm that most likely can be environmentally and economically transferred. The "mostly likely to succeed" candidate is supposed to be determined by the simulation model of Chapter 5. Luckily, this issue can be simplified from what we learned in Chapter 1. Most of the developing countries of our sample are located in tropical areas with somewhat similar economic conditions. Environmental and economic criteria can, therefore, be based upon this fact. As we discovered above, most of the aquaculture operations are concentrated in Southeast Asia which contains many developing nations. Hence, transferrence from one part of the developing world to another part seems most logical on the face of it.

Nonetheless a generally hospitable environment for production is not the sole consideration. The cultural acceptability of fish of a particular species must also be considered. And, by no means least, we are constrained by data availability. Economic data on aquaculture enterprises throughout the world are sparse. Bits and pieces of information must be put together in order to form a rudimentary profile of a "typical" aquaculture enterprise in any given country.

Biospheric Considerations. At least two of the biological characteristics of species have direct economic implications, the species' position in the food chain and dependency upon natural breeding. A species that is high on the food chain requires a high cost of feeding. Most aquaculture operations depend upon natural breeding. Although several species can be stocked both at temperate and hot tropical climates, the natural breeding of some species is very sensitive to variations in temperature. Thus we need to consider the environmental variables that determine the suitability of a species for culture. The important factors, among others, are: (1) water temperature; (2) salinity; (3) length of growing season; (4) tidal flushing; (5) PH factor; (6) rainfall; (7) sedimentation of soil (coefficient of hydraulic permeability); (8) photosynthetic activity; and (9) natural food supply.

Given the above environmental factors, we established criteria that relate, in an optimum way, the biological characteristics of species and environmental response. In our selection of species and/or plants for technology transfer, we used the following criteria: the species would (1) feed low on the food chain (needs little or no feed and/or fertilizer); (2) have a fairly wide temperature tolerance; (3) have limited breeding problems; (4) have a high growth rate; and (5) have an existing body of fairly well researched literature.

In view of the above discussion, prior acceptability of species in various nations makes a species a more positive candidate. Moreover, adaptability to small farm operations and high potential for foreign exchange would fulfill certain economic goals to a developing nation. Thus we added the following socioeconomic criteria: the species would (1) have established and profitable enterprises operating for some years somewhere; (2) have experienced use as protein in many countries; (3) have potentially high consumer acceptance; (4) be adaptable to small farm operations; (5) have labor employment potential; and (6) have a high potential for foreign exchange. Beyond these criteria, we also had—by necessity—to be guided by the availability of cost, earnings and factor input data.

The basic approach was to use Bardach *et al.* (1972) as the major source of descriptive material on aquaculture throughout the world, along with additional references and advice from numerous experts. This gave us a long list of species from which we could make a selection, using the criteria specified above.

Selected Species.   Table 2-4 shows the fish and plants selected for the simulation study along with their associated environmental characteristics. Notice that *carp, tilapia, prawns, oysters, milk-fish, mullet* and *mussels* are all species identified above as being the most extensively cultured. Indian carp, for example, feeds low in the food chain; has wide temperature tolerance; lives in both acid and alkaline water. Mussels need little feeding (in Thailand) and are cultured under very simple conditions. Tilapia are tropical, and widely

Table 2-4. Aquacultured Species Selected for Technological Transfer to Various Developing Countries: Some Environmental Characteristics

Species	Scientific Name	Water Temp. Req.	Salinity Req.	O_2 Req.	Acid-Base Req. (pH)	Other Env. Factors	Feeding Habits	Fertilizer	Eggs, Fry, & Where Obtained
1. Indian major Carps	a. *Catla catla* b. *Labeo rohita* c. *Cirrhinus mrigala*	>20°	slight tolerance of salinity	Needs >3 mg/1; 6-7 mg/1 optimal	7.3 – 8.4	Will not spawn in still waters; spawned in bunds in India.	filamentous algae & zooplankton; supplemented feeds (yeast, Vitamin B_{12})	Cow dung for zooplankton growth	Consistent success with pituitary extracts
2. Catfish (channel)	*Ictalurus punctetus*	21-24° for repro; 16-30° survival range; 28-30° optimal for growth	withstand slight salinity	2-3 ppm low threshold	7 – 9	Susceptibility to fungus	Specialized supplemental feeding	Discouraged, especially organic fertilizer	Pond and pen spawning
3. Walking Catfish	a. *Clarias batrachus* b. *Clarias macrocephalus*	25-32° for hatching	Withstand slight salinity	Not a factor	Not a factor	Shallow water for repro	Omniverous		Spawn in captivity. Also, pituitary injection
4. Tilapias	T. *mossambica* T. *nilotica* T. *maetochir*	12->30°	Very tolerant *mossambica* can be adapted to 40% S.W.	Tolerant of low O_2	No info found	Very rapid growth rates	Mostly phyloplankton eaters	Organic fertilizer used, also phosphates and lime	Breed naturally
5. Mullet	*Mugel cephalus*	3-35°	0-38 ppt	No info found	No info found		As adults Herbivores. Larvae are fed phyloplankton, rotifers, brine shrimp	Superphosphate and ammonium sulfate	Spawning induced by gonadotropin injection. Fry culture difficult.

Table 2-4. *(cont.)*

Species	Scientific Name	Water Temp. Req.	Salinity Req.	O₂ Req.	Acid-Base Req. (pH)	Other Env. Factors	Feeding Habits	Fertilizer	Eggs, Fry, & Where Obtained
6. Milkfish	*Chanos chanos*	>15° for optimal survival	0-140 ppt	Can tolerate Low D.O. No problem	pH >7 alkalinity >150 ppm; hardness >100 ppm; Ca >30 mg/l		Lab-lab (algae cultivation) Filamentous blue-greens and benthic diatoms are best	Best fertilizer is N-P-K, S-18-4	Wild stock source of fry
7. True eel	*Anguilla japonica*	>18° for repro	Raised in F.W. but tolerant of S.W.	High D.O. needed for optimal feeding	No info found prob. pH >7	Disease and cannibalism problems	Fed on Tubifex worms, chopped fish	Not used	Inshore migration of elvers main source of eels
8. Yellowtail	*Seriola quinqueradiate*	18-29° optimum for growth	16 ppt to S.W.	Must be >3 ppm	No info, prob. pH >7	Cannibalism if fry not graded. Cage culture vibrosis is common prob	Artificial feed for final phase of culture is 70% fish meal. Fry fed shrimp and white flesh fish	Not used (open sea water)	Wild stock source of fry
9. Shrimp	*Penaeus* spp.	25-29° optimal for spawning and growth	24-30 ppt optimal; 18-35 ppt can be survived for a few days by most species	80% of saturation optimal: 3.5 ppm is minimum. SINGLE MOST CRITICAL FACTOR!	No info	Cannibalism (losses up to 30% tolerated)	Ground clams, shrimp, minced fish	Not used (open sea water)	Post larvae collected from wild. Some artificial spawning.

		Temperature	Salinity	Oxygen	pH		Food		Seed/Spawning
10. Oysters	*Crassostrea* spp.	Spawning temp 20-25°, withstand wide fluctuations (15-30°)	23-28 ppt optimal	No info, probably high requirement	No info	Water must be changed completely & frequently by tides or current	Phytoplankton	Not used (open sea water)	Either collection of natural sources of spat, or hatchery production of seed oysters.
11. Mussels	*Mytilus* spp.	12-18° optimum	35 ppt optimum; heavy rainfall causing FW not a serious problem	No info, probably high requirement	No info	Water must be changed completely & frequently by tides or current	Phytoplankton	Not used (open sea water)	Seed mussels collected from natural beds then tied on rafts
12. Blue-green algae	*Nostoc commune*	Wide tolerance	No info	No info	No info		Not needed	Not used	Spores collected artificially. No info
13. Freshwater prawn	*Macrobrachium rosenbergi*	Incubate at 26-28°. Optimum for adults 26-28°	Adults - fresh to seawater Larvae - 8-22 ppt; optimum for adults 12-14 ppt.	Must be near saturation.	7.0 - 8.0	Many fishes can be polycultured with FW prawns	Fresh ripe fish eggs; also mollusks & earthworms. Incurable fungus if overfed.	Not used	Artificial spawning.
14. Rainbow trout	*Salmo pairdneri*	5-25° tolerance; 8-13° for hatching	Freshwater only	Must be near saturation for hatching, high for adults	>7.0	Carbonic acid, super-saturation of Nitrogen, and elimination of Ammonium from water important considerations	Artificial feed, with proteins 35-40%, CHO, 8-10% fat, vitamins A, D, B$_{12}$, etc.	No organics	Eggs artificially obtained

Primary Sources: Bardach, J.E., Ryther, J.H., and McLarney, W.H. (1972). Huet, M., and Timmerman, J.A. (1971).

cultured. However, some selected species violate some of the biological criteria, but were included where socioeconomic considerations dominated. For example true eel must be fed protein, but may offer some potential for foreign exchange.

We shall not try to justify each of our decisions on a species basis, but our research team (using the criteria above as general guides but realizing that some criteria are in conflict with others for certain species) felt that the species selected would be *a priori* high in transferability. And, we again remind the reader of the great restraint of economic data availability which is discussed in more detail in the next chapter.

In the main, aquaculture species on which to concentrate as economical sources of protein would be those species which are fairly popular as food, low on the food chain, hardy, easy to culture and fast growing. Such species as carp, catfish, tilapia, milkfish, mullet, and mussels exemplify ones that need to receive the largest concentration of aquaculture effort and attention. If this is done aquaculture probably will make an increasingly important contribution to world protein production.

REFERENCES

Bardach, John E., Ryther, John H., and McLarney, William O., *Aquaculture: The Farming and Husbandry of Freshwater and Marine Organisms*, New York: Wiley-Interscience, 1972.

Bell, Frederick W. *et al.*, "The Future of the World's Fishery Resources: Forecasts of Demand, Supply and Prices to the Year 2000 with a Discussion of Implications for Public Policy," U.S. Department of Commerce, National Marine Fisheries Service, Working Paper 71-1, December 1970.

Caton, Douglas D., Moss, Donovan D. and Urano, James A., "Improving Food and Nutrition Through Aquaculture in the Developing Countries," 5 March 1974 (unpublished).

Consultative Group on International Agricultural Research, Technical Advisory Committee, "Report of the TAC Working Group on Aquaculture," 7th meeting, Rome, 4-8 February 1974, Food and Agriculture Organization of the United Nations.

Davidson, Jack R., "Economics of Aquaculture Development," in Proceedings: Fourth National Sea Grant Conference, University of Wisconsin, Madison Sea Grant Publ. WIS-SG-72-112: 83-98. October 1971.

Delaney, Richard J. and Schmittou, Homer R., "Aquaculture Production Project, Philippines," World Bank Loan Proposal (unpublished).

Furukawa, A., "Outline of the Japanese Marine Aquaculture," Japan Fisheries Resource Conservation Association, Tokyo, 1971.

Huet, M., and Timmerman, J.A., *Textbook on Fish Culture—Breeding and Cultivation of Fish*, London: Fishing News Books, Ltd., 1971.

Ling, S.W., "A Review of the Status and Problems of Coastal Aquaculture in the Indo-Pacific Region," in Pillay, *ed. Coastal Aquaculture in the Indo-Pacific Region*, Rome, 1972.

Pillay, T.V.R., "The Role of Aquaculture in Fishery Development and Management," paper presented before the Technical Conference on Fishery Management and Development, Vancouver, 13-23 February 1973.

"Survey on Aquaculture in Southeast Asia," Report of the Japanese Survey Team, The Meeting of the Working Group of Experts on Aquaculture, Southeast Asian Fisheries Development Center, Manila, July 27-August 1, 1970.

Chapter Three

The Nature of Aquaculture Enterprises and Selected Models

THE NATURE OF GENERAL PRODUCTION ACTIVITY

The actual and potential production possibilities in a developing country can be best appraised by studying the economic, social, political and physical characteristics of the country and how they affect the levels of production. Production organization could be one of four different types: (1) Government, (2) Private Co-operative, (3) Commercial, and (4) Family-operated. Most of the developing nations listed in Chapter 1 have a large private enterprise system and hence most production of consumption goods takes place in the private sector. The literacy rate in these countries is low and the per capita GNP is also very low compared to the advanced economies. There is a large proportion of poor people living in the rural sector. The lack of proper infrastructure needed for industrial growth such as banking, insurance, transportation, technical skills, etc. forces the rural poor to engage in agricultural production activities which use land as the largest share of their productive inputs. The regional economic organization in these countries is such that urban centers are surrounded by rural areas that supply the food needed by urban people engaged in industrial and tertiary activities. Further, there may exist several functionally independent rural areas self-sufficient in their food and other basic needs.

THE FACTORS RESPONSIBLE FOR LOW LEVELS OF AQUACULTURE PRODUCTION

Given this general description, not surprisingly the total regional aquaculture production is often limited by the extent of regional demand, which in turn is restrained by the low income in the region. Even where excess demand exists in a region, often production increases are not forthcoming. Production can be increased either by employing more capital or adopting a new technology. But capital is usually scarce and new production technology relatively risky. Since most of the producers are either single family units or small-scale producers with small margins of profit they cannot undertake the risk of adopting a new technique unless its potential economic benefits are demonstrated beyond any doubt. Even though there exists a small number of rich individuals in the rural areas who could make the much needed investment in production, they often find it more remunerative to invest in real estate or in industrial or business enterprises in urban centers undergoing rapid urbanization.

Only a few of the developing nations engage in aquaculture to any substantial extent. Those that do are in Southeast Asia, a phenomenon explained by the history and sociology of the inhabitants. For example, although India is the second largest producer of aquaculture, most of its aquaculture takes place in West Bengal, a state of fish-loving people. Elsewhere in India, a large proportion of the population does not consume fish. In rural western Uttar Pradesh, a typical rural area in India, per capita monthly expenditure on fish *plus* meat and eggs is only about 1 percent of the total consumption expenditure (Singh, 1973). Religious, cultural and traditional forces limit both the horizons of rural investors and of rural consumers. Sometimes these limiting food and investor preferences may be regional, as in the Indian example. Sometimes the patterns are nationwide.

In Latin America, the technology of aquaculture is not as well developed as in Southeast Asia. In most Latin American countries, with the exception of Brazil and Mexico, the aquaculture technology is not extensively developed.

TYPICAL AQUACULTURE OPERATIONS

The relatively low levels of industrial output, capital and skills in developing nations give rise to aquaculture production of a labor-intensive nature. The population pressure on land also explains why most aquaculture enterprises are small in the size of land holdings. In addition, the dependence on land by the rural population and the institutional and demographic factors have contributed in some countries, such as India, to fragmentation of land holdings.

There are three types of aquaculture operations in Southeast Asia: (1) pond culture, (2) raft or cage culture, and (3) rice or paddy field culture. All three methods are used for fresh water rearing. Methods (1) and (2) are also adaptable to brackish water aquaculture, and Method (2) is used in the open ocean and estuarine areas. Feeding systems vary depending upon the species reared, but generally are of two types: (a) direct feeding of some kind of food that is low in price and high in protein for those species that can benefit from this kind of operation; and (b) fertilization of the aquatic environment through the addition of nitrates and phosphates so as to raise the level of natural food production on phytoplankton food chain. In some cases no artificial feeding or stimulation of aquatic foods is undertaken and the fish reared depend entirely upon natural foods, developed naturally.

In general the kind of culture (pond, cage, etc.), the type of water used (fresh, brackish, or open sea), the feeding system adopted (human food waste, prepared food, or fertilization) will depend on the relative economic costs and benefits under given environmental conditions. Common carp, for example, are most cheaply reared in fresh water ponds with some fertilization of the water to raise the algal level. However, if fresh water is in short supply, carp can be reared in brackish water. Some supplemental feeding of carp is often beneficial as they are omnivorous. The cost-benefit ratios among alternative species, the feeding levels, the feeding systems, the feeding formulations, the type of culture, and the type of water management depend upon a nation's cultural preferences, environmental constraints, and general economic conditions. Not surprisingly, then, the nature of aquaculture operations varies between countries, between the types of cultures, and between the species.

The pattern of local demand, the economic potential for exports, land ownership and land lease situations are some of the important economic factors determining the species cultured. For example, in India the states of Kerala, West Bengal and Orissa all have coastlines. Yet the pattern of local demand and export potential dictate that Kerala produce predominantly a brackish-water prawn species for export while Orissa and West Bengal produce a fresh water species, the Indian major carp, for local consumption. In Kerala the local demand for rice and coconut are very high, and the land utilization for fresh water ponds competes with land utilization for rice and other foodgrain production. The coastal land with greater amounts of salinity is unsuitable for rice cultivation, and hence it has relatively low market value. Kerala's shrimp culture consists of stocking the larvae and juvenile shrimp until they reach a marketable size. The low land rent and the low labor costs, the only input costs of such culturing, makes the shrimp culture a very profitable enterprise.

In Orissa and West Bengal there is a great local demand for Indian major carp which is a fresh water species. In some instances the carp culture is also carried out in brackish water when the salinity is low from rain water. Carp are also cultured in the estuarine waters of these states.

The environment can, of course, provide a constraint on the production of certain species. Eel and trout cannot be cultured in the hot tropical climates of India, Thailand, Burma, Sri Lanka, and Vietnam. They can, however, be cultured in the temperate climates of Taiwan and Japan. Moreover rainbow trout is commercially cultured in the Jammu and Kashmir States of India at high altitudes.

In the developing economies the aquaculture operations in current use are mostly based on historical tradition. As we stated above, most aquaculture operations in the developing world are undertaken by small individual farmers with either hired or family labor. Also, we noted that the population pressure on rural land and the pattern of local demand often determine the extent and the type of species cultured. Given the economic and environmental variables the type of aquaculture actually in use thus need not be the best possible technology.

There is hope for some breaks with tradition, however. In most of these countries there are fisheries departments and groups of fisheries scientists whose chief aim is to study the breeding of fish and their growth in confined waters. There exist, therefore, some aquaculture operations run by the government departments of fisheries. While these operations are mostly experimental, in some countries, for example in India and in Brazil, the experimental ponds run by the government function as the hatcheries supplying the fish fry and fingerlings by procuring them from various sources at the prevailing market price. For example, the Fish Seed Syndicate situated in Howrah near Calcutta (India) supplied fish fry to various state governments in 1973-74 for distribution to the fish farmers. Such government operations will have to expand in order to speed the growth of aquaculture.

THE ECONOMIC ASPECTS OF AQUACULTURE ENTERPRISES

Performing satisfactory economic analysis of aquaculture enterprises is made extremely difficult by the paucity of appropriate raw data. In this section we can nonetheless give a brief survey of the kinds of economic information available and describe the economic aspects of enterprises from such fragmentary data.

In addition to the aforementioned work of Bardach (1972) the Indo Pacific Fisheries Council (IPFC) with its headquarters at the FAO's regional office in Bangkok has undertaken the task of collecting economic data from the member countries. There are principally two IPFC data sources that were made available to us. One of them is the report of the first IPFC Working Group on the Economics of Aquaculture (1970). This report gives economic data by species for several countries. This source was used to study the problems of transferring the aquaculture technology that will be discussed in the next section. A second potential data source is the preliminary analysis of questionnaires (designed by T.V.R. Pillay) conducted in April 1973 by the IPFC Working Party on the Economics of Aquaculture. Unfortunately, in this analysis the species cultured were not identified, and therefore the data must be viewed as averages for aquaculture operations.[1]

Some studies suggest using the rate of return to capital as a measure of profitability of the enterprise. An example of such returns is provided by the preliminary IPFC analysis. The reported average profit per hectare on private farms in India is Rs 2466.5. The rate of return on investment varies from 3.1 percent to 143.6 percent whereas the short run market rate of interest is about 4 percent. The profitability of cooperative farms is much less and some of the Government farms incur losses. The degree of variability suggests that the profitability of the enterprise depends on the management practice followed and the species cultivated.

From the point of view of economic development and increasing the production of animal proteins in the developing nations it may well be preferable to choose an enterprise that yields a low rate of return to capital. Even though the private rate of return to capital could be low for a particular species and management type, other social goals may be met by such investments. For example, investment in land improvements and equipment for pond construction could be expected to generate additional employment in the rural sectors that have surplus agricultural labor. Aquaculture production would increase the amount of animal protein for domestic consumption. In some instances such as shrimp production in India, an aquaculture operation may generate foreign exchange earnings. If the social goals of a nation include lower unemployment, a higher level of animal protein production, and greater foreign exchange reserves, then the social rate of return to such capital may exceed the private rate of return.

---

1. Our research team requested access to the UNFAO file on the economics of aquaculture, but Dr. Pillay would not release the data.

Wherever there is a low private rate of return and/or possible high risks involved in these operations, there must be a considerable effort by the public sector to intervene and create incentives for investment in aquaculture. If such incentive measures cannot generate the required investment, the public sector should make the required investments. The low level of investment in aquaculture operations is mainly due to the lack of credit facilities for the fish farmer. Efforts should be made to make credit available to small scale aquaculture operations along with determined efforts by the credit granting agency to supervise the operations effectively.

Next we will consider models of particular aquaculture enterprises for possible transfer. One farm each is modeled for our selected species.

MODEL ENTERPRISES SELECTED FOR TRANSFER

Input-Parameter Simulation Data. Fourteen species were selected in Chapter 2 for the simulation study. In turn it is necessary to use (at least) fourteen model aquaculture farms in order to simulate the production of each species. The models are selected from various nations, depending upon the most reliable data available for each of the selected species. Because of our interest in small farms and the potential impact upon the income distribution, we chose small-farm data wherever possible.

In order to model each farm, input parameters are required. Because we assume that the exact farm could be transferred these parameters are constant across countries even though national factor prices vary (see Chapter 4). The selected data are presented in Table 3-1. Because of the special difficulties encountered in finding such data the complete citations for the sources used (except for those already referenced) for this table are given in Appendix 3-A. Next we will discuss the nature of each of the enterprise types.

Descriptions of Aquaculture Enterprise Models Selected. The following describes the general characteristics of the culture as well as qualifications that apply to our factor input data.

1. Indian Carp: Carp production takes place mostly in freshwater ponds. The average size of the farm is about 2 to 3 hectares split into about 10 ponds. The source of water in most of the ponds is rain water. This culture method does not use any feed. Rather, since this species is a plankton feeder, the ponds are usually fertilized. The fry and fingerlings stocked in these ponds mostly come from shallow river waters during the monsoon season.

Table 3-1. Factor Input Quantities for Aquaculture Products in Various Countries of the World

Species	Country of Origin	Source	Stocking (Number)	Feed (kg.)	Size of Farm (ha.)	Fertilizer (kgs.)	Labor (Man-years)
1. Indian Carp	India	Pillay	2,500,000	360	3.5	10,000	10.4
2. Channel Catfish	Georgia (U.S.A.)	Georgia	10,000	8,164.67	2.025	217.7	.075
3. Walking Catfish	Thailand	IPFC	375	1,779.94	.00152*	000.00	.06
4. Tilapia	Thailand	IPFC	60	67	.1075	4.0	.02
5. Mullet	Hong Kong	IPFC	25,000	80,906	5.0*	000.00	5.57
6. Milkfish	Philippines	Carand Darrah	1,200,000	68,038	100.00	907,185[1] 181,437[2]	6.58
7. True Eel	Taiwan	Shang Chen	120,000	472,500	1.0	000.00	2.44
8. Yellowtail	Japan	Pillay	3,110	112,379	.0126	000.00	1.057
9. Penaeus Shrimp	Thailand	Pillay	000.00	000.00	8.0	000.00	5.23
10. Oysters	Japan	Bardach & Others	000.00	000.00	1.0	000.00	1.0
11. Mussels	Thailand	Pillay	000.00	000.00	.16	000.00	3.64
12. Seaweed (Blue-Green Algae)	Africa (Chad)	Sorensen	000.00	000.00	4.0	453.59	6.00
13. Macrobrachium	Hawaii	Shang	924,500	249.12[3]	4.05	000.00	2.0
14. Rainbow Trout	Japan	IPFC	857,142	196,226	.36*	000.00	10.6

*Sea Water
1. 16-20-0 fertilizer
2. Chicken manure
3. Chicken feed
Sources: See Appendix 3-A.

2. Channel Catfish: The fish has very good growth characteristics, and provides a good boneless fillet. There are some problems in its culturing, however, which make it slightly more difficult than some species to cultivate. Also, the rearing method includes feeding the catfish a prepared food which is expensive. The specific operation modeled is based upon a composite of fresh water pond experiences in Georgia (U.S.A.) The scale of operation assumed is five acres.

3. Walking Catfish Culture: This species also is cultured in fresh water ponds. The average size of the farm is about three hectares split into about eight ponds. The source of water is either rain water or river water, often pumped in and out using a motor. Feed is used in the rearing ponds and the feed consists of trash fish and rice by-products. The ponds are stocked with fry and fingerlings that nature provides in shallow portions of river streams during the monsoon.

4. Tilapia: Even though tilapia are probably grown more universally than any other fish except carp, and can be reared in fresh, brackish or saltwater environments, surprisingly few data are available on their rearing. This may be because tilapia breed quite freely and there is a problem of excessive young so that they frequently are reared on a small scale or in a poly-culture. Certainly, however, the species does adapt especially well to the small farm environment and it is from this environment that data on its culture were collected. The stocking material can be purchased for a modest sum. A limited amount of feeding is usually undertaken and even smaller amounts of fertilizer purchased.

5. Grey Mullet: This species is cultured in brackish water ponds in several countries, including Mediterranean and Indo-Pacific countries. The fry and fingerlings are provided by natural spawning that takes place in estuaries and coastal tidal waters in late winter or spring. Mullet are tolerant to variations in temperature and salinity. The size of the brackish water ponds used for mullet culture varies from country to country. Mullets are cultured either alone or with Chinese Carp. When they are cultured alone they are intensively fed by rice bran and peanut cake (in the Indo-Pacific region).

6. Milkfish: In the Philippines, milkfish culture is the main aquaculture activity and provides some 95 percent of the country's supply of that species, which is highly favored in the market. The milkfish data are based upon a 100 hectare operation in the

Philippines, with a program of some supplemental feeding as well as the fertilization process normal in the rearing of that species. Fish fry are captured from wild stock, reared in ponds and sold in the market at about one kilogram or less in size. The large operation modeled reflects some economies of scale not possible for small farmers, but nonetheless is selected for two important reasons. First, the fisheries estate development program now going on in the Philippines will provide several thousand hectares of government developed rearing ponds for private lease to small scale farmers. Thus economies of scale will be made available to farmers who lease under this operation. Second, there is a data constraint. Data on a much smaller scale operation are available, but are based upon the less productive lab-lab method of production.

7. *True Eel:*   The enterprise is modeled on the basis of an assumed four hectares of land (three hectares of pond area) operation. A stocking rate of 120,000 seed eel per hectare was assumed with a 60 percent survival rate. The seed eels were assumed to have been gathered from the wild and sold to the eel farmer who feed trash fish on a 12:1 feed conversion rate. Output of the pond per hectare was taken to be 12,000 kg. per year.

8. *Yellowtail:*   This is cultured mostly in Japan in floating net cages, 35 to 100 square meters in area and three to six meters deep. This species is culturable only in temperate brackish water with temperature ranging between 18° to 29°C. This culture needs prepared food that includes fish meal. An average farmer spends about 50 percent of his operating expenses on the feed. The spawn is collected in late April or early May in coastal sea water.

9. *Penaeus Shrimp:*   All shrimp seed comes into the ponds with the initial impounded water and later with the subsequent flow of the incoming tides through pumping. There are ditches at least two meters wide along the embankment of the pond to provide shelter for shrimp. At present (for Thailand) no fertilizer or feed are employed in shrimp farming, but stocking is repeated through the culture period. Most shrimp farms harvest four to six crops a year.

10. *Oysters:*   The culturing of oysters can be done in a variety of ways but the most common form of intensive culturing is to rear the oysters on strings attached to floating rafts. This is the culture method taken as the model in the analysis. The cost of construction of the rafts, assuming 600 strings to the raft, each string 10 meters

long, and five rafts to the hectare is drawn from data collected during some experiments in Delaware Bay. The environmental data, production rates and prices are modeled on the basis of information obtained on the Inland Sea of Japan.

11. Mussels: Mussel spat (larvae) falls occur almost year round in the estuaries. Mussel growers stake bamboo poles into the ground at depths up to 10 meters to collect spat from December to March. The distance between poles is normally one meter. Bamboo stake traps are also used as collectors. Mussels of the first two spatfalls are usually harvested for use as duck feed, when they are about two months old. The third setting of spat is then allowed to grow to marketable size, which takes about eight to twelve months. This type of operation would be very suitable to small farmers.

12. Blue-Green Algae: This algae is widely used in China where it is grown in lagoons or brackish water pools. Little sophisticated technology is involved. Laborers use a hand rake and strainer to harvest the wet algae. The algae usually feeds on natural sewage outfall. This plant species can grow practically anywhere, except in very high or very low water temperatures.

13. Macrobrachium: Macrobrachium are included in the study because they appear to be easy to rear, using poultry starter as a feed, and stocking ponds with fry obtained from a central stock rearing operation. Macrobrachium are expensive to rear because of the feeding cost, but currently sell at high prices in export markets. In experiments in Hawaii macrobrachium reach a size of about eight ounces in one year, and are an extremely popular gourmet item. The cost or production per unit of protein generated, however, is high. Data on the macrobrachium culture are based upon information gained in a 10-acre operation in Hawaii.

14. Rainbow Trout: Rainbow trout are widely cultured throughout the world. They are usually cultured in freshwater lakes created by springs with low temperature waters. In Denmark and Japan where freshwater is scarce, however, trout are cultured in brackish or estuarine waters. The fry and fingerlings are usually produced in commercial hatcheries. This culture needs intensive feeding. Different countries have different feeding formulas. An optimum food is considered to be 60 percent protein, 10 percent carbohydrates, 25 percent fat, 5 percent minerals and a variety of vitamins.

Fish are not, of course, the only source of protein for developing nations. Aquaculture must compete with alternative sources. These alternative supplies as well as their prices are considered in the next chapter. Also, we will construct the factor prices that correspond to the production parameters given above.

Appendix 3-A

Simulation Data on Factor Input Parameters for Modeled Enterprises

The data sources for the factor input quantities of Table 3-1 are:

1. *Indian Carp:* T.V.R. Pillay, "The Role of Aquaculture in Fishery Development and Management," FAO Conference on Fisheries Management and Development, Vancouver, 1973 and Bardach (1972).

2. *Channel Catfish:* E. Evan Brown, M.G. LaPlante, and L.H. Covey, "A Synopsis of Catfish Farming," Univ. of Georgia College of Agriculture, September 1969, Bulletin 69.

3. *Walking Catfish:* IPFC (1970).

4. *Tilapia:* N.B. Jeffrey, "Progress on the Development of Fisheries in Northeast Brazil," Auburn Univ., 1972 and IPFC (1970). Also, see Brown, LaPlante and Covey under Channel Catfish.

5. *Mullet:* IPFC (1970) and Bardach (1972).

6. *Milkfish:* Philippines Fisheries Commission, Development Bank of the Philippines, Presidential Economic Staff, "Philippine Fisheries Industry Development" (A Loan Proposal to the World Bank), April 1972; and F.L. Carandang and L.B. Darrah, "Bangus Production Costs," Marketing Research Unit, National Food and Agricultural Council, Department of Agriculture and Resources, Quezon City, May 1973.

7. *True Eel:* Yung C. Shang, *Economic Aspects of Eel Farming in Taiwan*, Chinese-American Joint Commission on Rural Reconstruction, Fisheries Series No. 14, 1973, and T.P. Chen, "Eel Farming in Taiwan," August 1973, mimeo.

8. *Yellowtail:* Bardach (1972). Also, see Pillay under Indian Carp.

9. *Penaeus Shrimp:* Dept. of Fisheries, Ministry of Agriculture, Thailand, "The Result of Cost and Earning Survey on Shrimp Culture in Thailand," 1970; and Arporna Sribhibhadh, "Status and Problems of Coastal Aquaculture in Thailand," in T.V.R. Pillay *ed.*, *Coastal Aquaculture in the Indo-Pacific Region*, FAO, Fishing News Ltd., 1970.

10. *Oysters:* Bardach (1972); Don Maurer and Glenn Aprill, "Feasibility Study of Raft Culture in the Delaware Bay Area," Delaware River Basin Commission, October 1973; and A. Furakawa, "Outline of the Japanese Marine Culture," Japan Fisheries Resource Conservation Association, Tokyo, 1971.

11. *Mussels:* Scribhibhadh (1970). Also, see Penaeus Shrimp.

12. *Blue-Green Algae:* These data are derived from personal conversations with Dr. Philip Sorensen of Florida State University. Dr. Sorensen relied upon Dr. J. Leonard, Professeur à la Universite de Bruxelles, Belgique; and L. Newton, "Uses of Seaweed," London: Methuen and Co. Ltd., 1970.

13. *Macrobrachium (Prawns):* Yung C. Shang, "Economic Feasibility of Fresh Water Prawn Farming in Hawaii," Economic Research Center, University of Hawaii, June 1972.

14. *Rainbow Trout:* IPFC (1970) and Bardach (1972).

OTHER REFERENCE

Singh, Balvir, "The Effects of Household Composition on Its Consumption Pattern," *Sankhya*, Series B, Volume 35, Part 2, 1973, pp. 207-226.

�֎ *Chapter Four*

1970 National Protein Supplies and Costs

INTRODUCTION

The purpose of this chapter is to provide the sources and the explanations of data derivation for the simulation study. The data include geodemographic production input variables, production output variables, and model-firm revenue production data. All in all 142 general demographic and economic variables were collected or constructed from various data for the 90-nation sample. Not all the variables entered the final analysis, but may be useful for further work in this area.[1]

The geodemographic variables include population, agricultural labor force, urban population shares, land areas of different types (arable, irrigated, meadows and pastures, agricultural). The aggregate economic variables include per capita GDP, per capita consumption expenditures, exchange rates and an index number of food production. Production input variables include fertilizer production and consumption by type (phosphate, potash, nitrogen), numbers of agricultural tractors, and fertilizer import values and quantities.

Among the most difficult data to develop were input prices. Because only operating-cost data are available at the firm level, the rental rate on capital is calculated as a residual. The cost of labor (agricultural wage rate) and the rental value of land had to be constructed. Protein production cost for each major source of protein was calculated for each nation in both raw and qualitatively

---

1. A complete listing of the economic and demographic data by sources is available from the authors.

adjusted terms. The details of how these data are developed appear below.

Production output variables include total metric tonnage of the most important protein sources in each nation, the supply of the same food commodities in grams per day, and the supply of calories per day from the same sources.

NET FOOD SUPPLY VARIABLES (QUANTITIES, CALORIES AND PROTEINS)

The general sources used in the development of these data are:

FAO Production Yearbook, 1972, Food and Agricultural Organization of the United Nations (Rome, 1972); and

FAO Food Balance Sheets, 1964-66, Food and Agricultural Organization of the United Nations (Rome, 1971).

Protein production data were collected for the year 1970 where possible, the most recent year for which a data base can be built. Where the 1970 observation is not available, the otherwise most recent observation is used. The 1970 production data are not broken down into as small categories as earlier data. However, the U.N. *Food Balance Sheets* gives a breakdown of per capita food consumption by commodity for each country. In order to obtain the degree of *detail* available only in the earlier yearbooks, we assume in our study that *shares* of specific commodities within food groups were the same in 1970 as when the more complete 1964-66 budget study was conducted. The more specific *Food Balance Sheets'* commodities and the more general categories from which they were derived are:

1. Cereals:
   a. Wheat
   b. Rice
   c. Coarse grains
2. Pulses, nuts and seeds:
   a. Pulses
   b. Nuts and Seeds
3. Meat:
   a. Beef and Veal
   b. Sheep and Lamb
   c. Pig meat
   d. Poultry

4. Fish:[2]
   a. Fin-fish
   b. Shellfish

Through the use of this modification, we are able to provide data on quantities per capita per day, calories per capita per day, protein per capita per day, and price per kilogram for the following eighteen foods:

1. Wheat	10. Beef and Veal
2. Rice	11. Sheep and Lamb
3. Coarse Grains	12. Pork
4. Potatoes and Starchy Foods	13. Poultry
5. Sugar	14. Eggs
6. Pulses	15. Fin-Fish
7. Nuts and Seeds	16. Shellfish
8. Vegetables	17. Milk
9. Fruits	18. Fats and Oils

Data sheets for each country have been maintained that list irregularities in the data as well as U.N. footnotes on the data. Among these foods sugar and fats and oils contain no protein. Moreover, the following foods are insignificant sources of protein in the countries studied: potatoes and starchy foods; pulses; nuts and seeds; eggs; and milk. Although all these foods were considered, they are not included in the final protein cost simulation.

FOOD PRICE VARIABLES

The general source of all food price data is the *FAO Production Yearbook* (1972). Where a *specific country price* is not available, an average unweighted price of the major nation producers (usually producer and otherwise wholesale prices) of that region is used as a proxy. If no country *or* regional price is available for a commodity, the average unweighted price of the major *world* producers of that commodity is used.

In virtually all cases where more than one commodity appears in a commodity group (e.g., pulses, beans, peas, etc.) the price of the commodity of greater output for that country is used for the price calculation. Records for the exceptions to this such as the average world vegetables or fruit prices were kept in our data sheets. Also, the special explanations for prices in five of our twelve developed

---

2. The fish category is almost exclusively fish catch from wild stocks.

countries appear on our records and can be made available. These nations are the United States, Canada, Australia, New Zealand and Japan.

All commodity prices are in U.S. cents per kilogram. However, because foreign exchange rates are collected, conversions from one currency unit value to another can be made.

The major U.N. classes provide an initial classification scheme for the price data. The U.N. classes and regions are:

Class I: Developed market areas

    Region A: Northern America
    Region B: Western Europe
    Region C: Oceania

Class II: Developing market economies

    Region A: Africa (including South Africa)
    Region B: Latin America
    Region C: Near East, Africa, Asia (including Israel)
    Region D: Far East (including Japan)

The U.N. Class III group of centrally planned economies are assigned to the above geographical regions (e.g., China appears in the "Far East": II; Region D). The price data for "Class II, Region B: Latin America" are presented in Appendix 4-A. The data for this region are illustrative of the price data base and also pertain to the area of "most probable transfer."[3]

PRICES OF THE FACTORS OF PRODUCTION

The development of price data for the factors of production was especially difficult. But such prices are required for the simulation costs in nations to which aquacultural technology is transferred (Chapter 5).

(1) The Price of Labor—Agricultural wage rates are reported by UNFAO for part of the sample. However, one would expect a high correlation between such wage rates and per capita incomes and therefore, per capita consumption expenditures. Nonetheless, the correlation between these reported wage rates and income per capita

---

3. Food commodity price data for all other regions including any special notes (as referenced above), are available from the authors.

variables for a sample of 30 developing nations was found to be extremely low ($R^2 = .08$). Therefore, we constructed an improved agricultural wage rate for our ninety nations. Our methodology follows.

The major food staple in each nation was identified. Then, per capita production of this staple was multiplied by the farm price of that commodity in the nation. The share of the total food budget represented by cereals, starchy roots and pulses declines as per capita income rises, the share spent for livestock products, vegetables and fruits increases, and the share represented by sugar, fats and oils rises through per capita income wages to about $1000 and declines thereafter. These relations have been estimated by UNFAO. Also, the proportion of total expenditures that go for food decline with rising per capita income. This relation has been reported by Burk and Ezekiel (1967).

These food share trends were utilized in estimating agricultural wage rates. The reciprocal of the staple share of the food budget was multiplied by the reciprocal of the food-expenditures share in order to derive a combined multiplier. In turn, this multiplier was applied to our per capita staple value in order to expand the value of staple production to the size of the total per capita budget. This gives us a per capita budget that is in agricultural-staple value. We assume that this is the agricultural wage rate.

The budget shares are reported in Table 4-1, and the food-expenditures shares with their associated multipliers appear in Table 4-2. The final multiplier and the estimated wage rate (at annual rates) are provided in Appendix 4-B.

These wage-rate estimates are deduced from the following assumptions: (1) crop value estimates can be derived from the reported crop prices; (2) farm workers labor a minimum of six days per week; (3) the extended family size in developing nations is 6.5 members; and (4) this family "subsistence" wage can be adjusted by the multiplier for disposable income and budget shares.

Three exceptions to the above method were employed. (1) In a few higher-income developing nations a livestock product is the main food budget item, and the initial multiplier is derived from that row of our Table 4-1. (2) In extremely low-income areas very large quantities of starch are consumed on site and the use of a market price greatly overvalues this production. In such cases we assumed that the staple should be valued at an opportunity cost near zero. Arbitrarily we assigned a value of one U.S. cent per gram to such protein. Most of these nations are located in Africa. (They are footnoted in Appendix 4-B.) (3) Agricultural wages reported by

Table 4-1. Major Groups of Commodities as a Percent of Total Food Budget (in U.S. Dollars)

GDP per capita	50	100	150	200	300	400	500	750	1000	1500	2000	3000	4000	5000
Cereals, Starchy Roots and Pulses	.76	.71	.64	.61	.60	.55	.51	.46	.42	.36	.32	.27	.26	.26
Livestock Products, Vegetables, and Fruits	.10	.13	.15	.20	.16	.18	.20	.24	.25	.34	.37	.44	.49	.49
Sugar, Fats, and Oils	.14	.17	.21	.20	.24	.27	.29	.30	.33	.30	.21	.29	.25	.25
Multiplier	1.32	1.41	1.56	1.64	1.67	1.82	1.96	2.17	2.38	2.78	3.13	3.70	3.85	3.85

Source: *FAO Agricultural Commodities Projections for 1975 and 1985 Volume 1*, 1957, p. 39.

Table 4-2. **Proportion of Total Expenditure for Food by Income Levels**

Per Capita Income	Per Cent for Food	Multiplier
$ 20 - 30	70	1.43
$100 - 200	50	2.00
$500	35	2.86
$1,000	25	4.00
$2,000+	20	5.00

Source: Marguerite C. Burk and Mordecai Ezekiel, 1967, p. 341.

official sources were used for all the developed nations except Australia.

As a check on the reliability of our estimated wage rates compared with reported agricultural wages, we again regressed per capita consumption in each nation against our 90 (now estimated) developing-nation wages and those (12) reported by the developed nations. The result of the regression which includes 102 observations is:

$$\text{Per Capita Consumption} = 54.35 + \quad .660 \text{ (Est. Annual Wage}$$
$$t\text{-values} \qquad (2.07) \quad (22.81) \quad \text{Rate)}$$
$$R^2 = .84.$$

(2) **The Rental Rate of Land**—The rental value of agricultural land was calculated as a residual. We assumed that the only "capital" of major importance used in developing nation agriculture is fertilizer. Then, the following steps were taken to arrive at our residual for each nation. (1) The value of agricultural output per hectare of agricultural land was calculated. (2) The number of agricultural workers per hectare was calculated. (3) The quantity of total fertilizer consumed per hectare was calculated. (4) The above quantities of inputs per hectare (2 and 3) were multiplied by their respective prices (wages as estimated above). (5) All dollar values were "normalized" into average annual rates. (6) The total of average labor and average fertilizer costs was subtracted from the value of agricultural output per hectare in order to derive the rental value of land (per hectare). These estimates of rental land values (as well as the fertilizer price used) are reported in Appendix 4-B.

(3) **Costs of Other Production Inputs**—Other factor inputs include feed (for certain species), fertilizer (for certain species) and fingerlings. Macrobrachium use poultry starter as a feed, a product whose

price varies with that of coarse grains. The other selected species that were given feed (Indian Carp, Channel Catfish, Walking Catfish, Tilapia, Mullet, Milkfish, True Eel, Yellowtail and Rainbow Trout) were fed fishmeal. Using the 1970 U.S. poultry starter as a base price, the prices for other nations were calculated in line with U.S.-other nation coarse grain price differentials. The 1970 world fishmeal price (a very competitive price) of $.04 per kilogram was adopted as the fishmeal feed cost for all nations. These prices are reported in the first table of Appendix 4-C.

Fingerlings—that is, juvenile catfish—are required in the culturing of Mullet, Walking Catfish, Yellowtail, Eel, Carp and Tilapia. The price of fingerlings is determined in direct proportion to the unit prices of the species in question. Thus the fingerlings prices in each nation were assumed to bear the same relationship to the adult market price per unit as in nations in which culturing of the species is an ongoing process. The fingerlings prices calculated are reported in the second table of Appendix 4-C.

INITIAL AQUACULTURE COST MODELS

The detailed descriptions of selected aquaculture enterprise models were given in Chapter 3. The quantities of inputs for each of the 14 species modeled were provided in Table 3-1.

From the same sources we collected data on output (in kilograms) and the producers' selling price. Profits (or capital rental rates) were not available. Since for simulation purposes we assume that each species will be produced with the quantity of inputs used by the models, it is necessary to calculate unidentified costs of production which includes profits.

Table 4-3 gives production levels, product prices, and therefore revenue for the modeled enterprises. We calculated operating costs by multiplying the known inputs by their known prices or (where not available) the input prices estimated for the nation in which the farm is located. Then these known operating costs were summed. The final column of Table 4-3 shows the share of total revenue not accounted for by known operating costs. This figure is used as a "mark-up" for all the simulated costs in various nations. It can be viewed as a combination of profits and other unidentified production costs.

The prices of these factors of production and other production inputs are used in the national cost of aquacultural production simulations of Chapter 6.

Table 4-3. Total Revenue and Identified Operating Costs of Selected Aquacultured Species

Species	Product Price Per kg. ($U.S.)	Production (in kgs.)	Revenue ($U.S.)	Sum of Identified Operating Costs ($U.S.)	Percent of Revenue Not Identified
1. Indian Carp	$1.20	4,866.00	$ 58,400.00	$22,550.15	63.4%
2. Channel Catfish	1.10	4,082	4,500.00	1,749.10	61.1
3. Walking Catfish	0.85	305.88	260.00	112.41	56.8
4. Tilapia	0.38	37.00	14.00	11.17	20.2
5. Mullet	0.03	516,667.00	18,600.00	36.143.50	-94.3
6. Milkfish	0.52	150,000.00	77,922.00	27,829.00	64.3
7. True Eel	5.00	12,000.00	60,000.00	49,567.00	17.4
8. Yellowtail	2.79	3,110.00	8,673.00	6,560.20	24.4
9. Penaeus Shrimp	0.80	10,274.00	8,208.95	875.70	89.3
10. Oysters	0.16	90,720.00	14,243.04	1,788.00	87.4
11. Mussels	0.41	7,073	2,900.00	389.80	86.6
12. Seaweed	0.23	18,869.	4,340.00	198.88	95.4
13. Macrobracium	6.61	13,616.00	90,000.00	19,403.63	78.4
14. Rainbow Trout	0.38	491,147	186,636.00	24,045.00	87.1

Source: See Appendix 3-A in Chapter 3.

COST OF PROTEIN BY TYPE

To serve as a potential protein source, cultured fish must compete with all other protein sources, including wild stock fish. Therefore, we had to derive estimates of the unit costs of these competing sources.

(1) **Raw Protein per Capita per Day Rankings**—The raw quantities of protein by type (in grams per capita per day) vary, of course, by nation. In order to identify protein dependencies, all nations were ranked by per capita per day protein sources. The protein types are those listed above. The major rankings are provided in Tables 4-4 through 4-8. Country names are abbreviated (as in our computer output).

The total raw protein rankings for 1970 appear in Table 4-4, and the rankings for animal protein appear in Table 4-5. The grams per capita per day represent edible protein and is not adjusted for quality (based upon amino acid pattern). Such countries (from our sample) as New Zealand, Argentina, France, the U.S.A., Iceland and Canada are at or near the top of the list while nations like Zaire, the Congo, Angola, Tanzania, Indonesia, Ghana, El Salvador, Bolivia and Haiti appear at or near the bottom. Many of those nations that rank low in total protein depend upon starches as their main source of protein. This is revealed by comparing the high rankers in Table 4-6 with the low ones of Tables 4-4 and 4-5. An improvement in the diets of the populations of such nations as the Ivory Coast, Ghana, Gabon, Nigeria, Dahomey, Zaire, the Congo, Cameroon and Uganda would require the replacement of their starch dependence by higher-quality protein sources. (The rankings for all other protein sectors were also derived by the authors.)

As Tables 4-7 and 4-8 belie, fish, especially shellfish, are not dominant sources of protein around the world. The absolute values of raw protein figures are nonetheless somewhat misleading. Adjustment for the quality of protein raises the importance of fish protein relative to a number of alternative sources in a number of nations. For example, the share of raw world protein derived from wheat in 1970 was about 30 percent and the share from fish, about 5.5 percent. Adjusted qualitatively this wheat share drops to about 23 percent and the fish share rises to roughly 7.5 percent.

(2) **The Nature of Protein Deficiency**—Obviously, if aquaculture is to serve to reduce the world food problem, it must do so as a source of high quality protein rather than as a source of simply more

Table 4-4. Raw Protein, Total

Rank	Country	Grams per Capita per Day	Rank	Country	Grams per Capita per Day
1	New Zeal	108.4	52	Tunisia	62.9
2	Argentina	104.7	53	Cuba	62.8
3	France	102.6	54	Cos Rica	62.0
4	Greece	98.9	55	Nigeria	59.9
5	U.S.A.	98.6	56	Venezuel	59.7
6	Hungary	97.9	57	Panama	59.2
7	Iceland	96.8	58	Ivory Co	59.1
8	Canada	95.8	59	Sudan	58.9
9	Poland	93.2	60	Cameroon	58.9
10	Yugoslav	91.6	61	Iraq	57.8
11	Israel	91.5	62	Khmer	57.7
12	Austrailia	91.5	63	Morocco	57.7
13	Uruguay	90.8	64	Yem Ar R	57.5
14	Denmark	88.5	65	China	57.2
15	Italy	87.9	66	Sua Arab	56.2
16	Romania	87.0	67	Uganda	55.9
17	U.K.	86.8	68	Algeria	55.7
18	Czechos	83.3	69	Iran	55.2
19	Germ F R	83.0	70	Honduras	55.0
20	Norway	81.9	71	Banglade	54.9
21	Portugal	81.8	72	Pakistan	54.9
22	Spain	79.9	73	Peru	54.6
23	Egypt	79.9	74	Philippi	53.2
24	Chad	78.4	75	Dahomey	52.2
25	Turkey	77.9	76	Nepal	52.0
26	Cyprus	77.9	77	Madagasc	51.2
27	Korea D R	77.8	78	Gabon	51.0
28	Niger	77.5	79	Thailand	50.5
29	S. Africa	77.0	80	Guatemal	50.5
30	Japan	76.9	81	Dom Rep	50.1
31	Rhodesia	73.2	82	Columbia	50.0
32	Korea R	72.4	83	India	49.4
33	Albania	71.3	84	Silcone	49.2
34	Upper Vo	70.3	85	Sri Lank	49.1
35	Lebanon	69.9	86	Malaysia	49.1
36	Zambia	69.4	87	V.Nam R	48.6
37	Syr Ar R	69.2	88	Haiti	47.0
38	Mali	68.4	89	Bolivia	45.8
39	China Ta	68.2	90	Ecuador	45.5
40	Kenya	68.0	91	Guinea	45.4
41	Brazil	66.8	92	El Salva	45.2
42	Ethiopia	66.3	93	V.Nam DR	44.8
43	Mexico	66.3	94	Laos	44.6

Table 4-4.   *(cont.)*

Rank	Country	Grams per Capita per Day	Rank	Country	Grams per Capita per Day
44	Lib Ar R	66.1	95	Burma	44.1
45	Chili	65.9	96	Ghana	43.0
46	Paraguay	65.4	97	Indonesa	42.8
47	Afghanis	65.4	98	Tanzani	42.5
48	Hong Kong	65.0	99	Mozambia	40.4
49	Jordon	64.8	100	Angola	39.9
50	Nicargua	63.2	101	Congo	39.8
51	Malawi	63.1	102	Zaire	32.7

calories. Thus we need to say something about the importance of protein. Until recently there was a much wider distinction associated with the terms "undernourishment" and "malnourishment." Undernourishment is widely used to represent the result of a shortage in total energy intake (measured in kilocalories) such that an individual cannot maintain normal activity without a loss in weight. Malnutrition is taken to mean the symptoms which develop from the lack or deficiency of one or more basic nutrients. The nutrient of importance, while usually protein, can also be one of the many vitamins or minerals.

The interrelationship between the symptoms and causes of these two conditions (undernutrition and malnutrition) has been discovered in recent years. Therefore, the two are now often seen as going together and are referred to as "protein-calorie malnutrition" (or protein-energy malnutrition). This condition also can be complicated by a number of diseases that can attack the body while in a weakened state.

Because infants and preschool children have less control over what they eat and because their ability to consume is limited, they are the groups most likely to suffer protein deficiencies without an accompanying energy deficiency. They are also the most likely to be afflicted with protein-calorie malnutrition.

Marasmus and kwashiorkor are the names given to the most severe clinical varieties of what is known in medical literature as the syndrome of protein-calorie malnutrition. Marasmus, caused by a gross shortage of both calories and protein, is characterized by shrunken features and gross physical retardation of the child. Kwashiorkor is brought on by a shortage of protein. The children often pictured with bloated bellies and glassy stares are victims of kwashiorkor.

Often because the effects of nutritional deficiencies can be studied

Table 4-5.   Raw Animal Protein, Total

Rank	Country	Grams per Capita per Day	Rank	Country	Grams per Capita per Day
1	New Zeal	73.9	52	Mali	15.0
2	Iceland	73.2	53	Turkey	14.8
3	U.S.A.	71.5	54	Malaysia	14.7
4	Australi	68.2	55	Rhodesia	14.4
5	Canada	64.0	56	Mexico	14.2
6	France	64.0	57	Chad	13.8
7	Uruguay	62.8	58	Iraq	13.6
8	Argentin	62.3	59	Ivory Co.	12.9
9	Denmark	57.9	60	Guatemal	12.7
10	Germ FR	54.5	61	Madagasc	12.6
11	U.K.	53.4	62	Niger	12.5
12	Norway	51.0	63	El Salva	12.5
13	Israel	44.3	64	Thailand	12.3
14	Hungary	43.4	65	Bolivia	12.1
15	Greece	43.0	66	Iran	11.7
16	Poland	42.6	67	Syr Ar R	11.7
17	Czechos	38.7	68	Jordon	11.3
18	Italy	38.3	69	V Nam R	11.1
19	Portugal	34.6	70	Ethiopia	11.0
20	Spain	34.6	71	Tunisia	10.9
21	Hong Kon	32.3	72	Cameroon	10.8
22	Japan	31.8	73	Zambia	10.7
23	Paraguay	29.0	74	Egypt	10.5
24	S. Africa	28.3	75	Yem Ar R	10.1
25	Chili	28.0	76	Morocco	10.0
26	Cuba	27.6	77	Pakistan	9.9
27	Gabon	27.2	78	Banglade	9.9
28	Cos Rica	26.8	79	Sua Arab	9.5
29	Panama	26.3	80	Angola	9.4
30	Venezuel	26.2	81	Korea R	9.4
31	Romania	26.2	82	Burma	9.2
32	Columbia	25.5	83	Si Leone	9.1
33	Cyprus	24.9	84	V.Nam D R	9.0
34	Yugoslav	23.1	85	Zaire	8.9
35	Dom Rep	21.5	86	Sri Lank	8.5
36	Brazil	21.4	87	Nigeria	8.4
37	Albania	21.2	88	Dahomey	8.0
38	China Ta	20.9	89	China Ma	7.9
39	Nicaragu	20.7	90	Afghanis	7.8
40	Philippi	20.6	91	Ghana	7.3
41	Lebanon	20.4	92	Korea D P	7.1
42	Lib Ar R	19.7	93	Nepal	6.7
43	Khmer	19.1	94	Algeria	6.6

Table 4-5. *(cont.)*

Rank	Country	Grams per Capita per Day	Rank	Country	Grams per Capita per Day
44	Peru	19.1	95	Guinea	6.0
45	Sudan	18.7	96	India	5.6
46	Congo	16.0	97	Upper Vo	5.3
47	Kenya	15.9	98	Malawi	5.3
48	Honduras	15.5	99	Laos	5.2
49	Tanzani	15.4	100	Indonesa	5.2
50	Ecuador	15.1	101	Haiti	4.7
51	Uganda	15.1	102	Mozambiq	4.6

under isolated conditions with test animals only, rather than with humans, insufficient proof exists to draw definitive conclusions concerning direct relationships between deficiencies and physical and mental subnormalities. This is the case when discussing the effects on physical growth and stature. It is believed, however, that nutrient deficiencies during periods of growth cause a permanent reduction in ultimate size in humans as it does in animals.

There are many documented cases that evidence an association between malnutrition in preschool children and low levels of mental performance. Studies also have correlated protein-calorie malnutri-

Table 4-6.   Raw Starchy Protein

Rank	Country	Grams per Capita per Day
1	Ivory Co	14.8
2	Ghana	13.7
3	Gabon	13.0
4	Nigeria	12.7
5	Dahomey	10.9
6	Zaire	9.4
7	Congo	9.4
8	Cameroon	9.3
9	Uganda	9.2
10	Angola	7.5
11	Mozambia	7.5
12	Portugal	6.6
13	Bolivia	6.2
14	Peru	6.1
15	Spain	6.0
16	Poland	5.9
17	Paraguay	5.9

Table 4-7. Raw Protein, Fin-Fish*

Rank	Country	Grams per Capita per Day	Rank	Country	Grams Per Capita per Day
1	Iceland	22.3	29	Dahomey	3.8
2	Khmer	16.0	30	Angola	3.8
3	Japan	16.0	31	Uganda	3.8
4	Portugal	14.3	32	Korea R.	3.6
5	Congo	11.3	33	Tanzani	3.6
6	Hong Kon	9.6	34	Mozambiq	3.6
7	Philippi	9.5	35	V. Nam D.R.	3.5
8	China Ta.	9.2	36	Zaire	3.5
9	Denmark	9.0	37	Indonesa	3.5
10	Norway	8.9	38	Panama	3.5
11	Gabon	7.9	39	S. Africa	3.5
12	V. Nam R.	7.0	40	U.S.A.	3.4
13	Spain	6.3	41	Chil	3.3
14	Si Leone	6.0	42	Italy	3.3
15	Greece	5.8	43	Dom. Rep.	3.2
16	Chad	5.7	44	Canada	3.1
17	Cameroon	5.6	45	Venezuel	3.1
18	Ivory Co	5.6	46	Poland	3.0
19	Malaysia	5.4	47	Cuba	3.0
20	Thailand	5.4	48	Australi	2.9
21	Sri Lank	5.2	49	Israel	2.7
22	France	5.2	50	Germ. F.R.	2.6
23	Zambia	4.7	51	China Ma	2.4
24	Burma	4.5	52	New Zeal	2.2
25	Peru	4.4	53	Madagasc	2.1
26	Ghana	4.3	54	Czechos	2.0
27	Mali	4.1	55	Brazil	2.0
28	U.K.	4.0			

*All populations consuming less than 2.0 grams per capita per day are excluded.

tion with depressed learning skills. The degree of nutrient and energy deficiencies, the timing of these deficiencies and the duration of their occurrence seem to determine the extent and permanence of the retardation of mental capabilities.

Some nutrition experts believe that protein-calorie deficiencies and subnormal mental functioning can be linked both directly and indirectly. Not only does the condition cause direct damage to the central nervous system, but it also has the following indirect effects: loss of learning time, interference with learning during critical periods of development, and changes in motivation and personality.

Table 4-8.   Raw Protein, Shellfish*

Rank	Country	Grams per Capita per Day
1	Panama	2.0
2	Dahomey	1.0
3	Nicaragu	.9
4	Korea R.	.9
5	China Ta.	.8
6	Malaysia	.8
7	Thailand	.8
8	New Zeal	.8
9	Australi	.6
10	CosRica	.5
11	Philippi	.4
12	Madagasc	.4
13	Chili	.4
14	Honduras	.3
15	Burma	.3
16	Portugal	.3
17	S. Africa	.3
18	Tanzani	.3
19	Cuba	.2
20	Ecuador	.2
21	Guatemal	.2
22	V. Nam R	.2
23	Brazil	.2
24	Venezuel	.1
25	Mexico	.1
26	India	.1
27	Argentin	.1
28	El Salva	.1
29	Iran	.1
30	Sua Arab	.1
31	Columbia	.1
32	Uruguay	.1
33	Sri Lank	.1
34	Peru	.1

*All populations consuming less than .1 grams per capita per day are excluded.

The direct physiological causes of malnutrition are many and varied, but for most economic purposes inadequate nutrition is a function of poverty. It is the driving force which binds together many other related problems characteristic of millions in the developing world with similar socioeconomic development patterns. Thus, most but not all of those suffering from malnutrition live in the developing regions of the world. According to the Food and Agricultural Organization (FAO) of the United Nations about 20 to 30 percent of all those living in the Near East, the Far East (excluding Communist economies) and Africa are in some sense suffering from inadequate nutrition.

From the foregoing clearly we need to use units of measurement other than simply "tons" of food. It is widely known that a pound of potatoes is not equivalent to a pound of beef in nutritional value. Yet much agricultural data are analyzed in aggregated forms that would assume so.

To begin with there is invariably some waste involved between the production site and that of consumption. Beyond this, one must differentiate between the efficiency of protein utilization and protein quality. Efficiency of protein utilization encompasses protein quality—an attribute of the protein itself and its amino acids.

Quality is a measure of the efficiency with which protein can be used for maintenance or growth. It is basically a measure of the amino acid balance. All proteins are made up of specific combinations of amino acids. The human body requires a minimum daily intake of at least eight essential amino acids. These eight amino acids cannot be synthesized by the human body. Protein efficiency varies among foods because of the different *patterns* of these eight amino acids. The quality of the protein in any type of food is constrained by the essential amino acid of the smallest relative amount contained in that food. A number of methods which attempt to measure this amino acid balance, and thus the protein quality of foods, have been devised (net protein utilization, protein efficiency ratio, biological value, and chemical score, among others). The method we employ is that of a chemical score. For practical purposes the protein of the whole egg, is used as a standard score. The values derived from any chemical score will depend on the scoring pattern used. The scoring pattern used in this study is the provisional one of Table 4-9.

The amino acid score of a protein or mixture of proteins is calculated by the following equation:

$$\text{Amino acid score} = \frac{\text{mg of amino acid in lg of test protein}}{\text{mg of amino acid in reference pattern}} \times 100$$

The lowest score obtained from any of the eight essential amino acids is then the chemical score for that protein. That specific amino acid is considered the "most limiting amino acid" for the protein under study. This score can be taken as an approximation to the probable efficiency of utilization of the test protein. The chemical

Table 4-9. Provisional Amino Acid Scoring Pattern

	Suggested Level	
Amino Acid	*mg. per g. Protein*	*mg. per g. Nitrogen*
Isoleucine	40	250
Leucine	70	440
Lysine	55	340
Methionine	35	220
Phenylalanine	60	380
Threonine	40	250
Tryptophan	10	65
Valine	50	310
Total	360	2255

scores of some common protein sources are given in Table 4-10. The scores used for our 14 selected species are given in Chapter 5.

Now, we are prepared to explain the adjustments-to-cost that were made for our simulation study.

(3) Cost of Raw Protein—For simplicity we assumed perfect competition as deciding producer food prices so that the producer (or occasionally wholesale) costs of the food commodities per kilogram are equal to the prices illustrated in Appendix 4-A.

The output quantity of the food commodity is not, of course, equal to the number of grams of protein derived from it. Because of

Table 4-10. Protein-Efficiency Conversions

Product	*Protein Score*
Milk	78
Egg	100
Beef	83
Pork	36
Mutton	94
Fish	75
Poultry	95
Rice	72
Corn Meal	42

Source: *Protein Requirements*, Report of the Food and Agricultural Organization of the United Nations, 1957.

wastage (bones, entrails, etc.) and the presence of non-protein substances such as fat, the raw protein content varies by type of food. In order to derive the cost per kilogram of protein, the following steps were therefore taken.

1. The conversion ratio of raw weight of the commodity to protein weight was recorded.

2. All commodity prices were multiplied by the reciprocal of this protein conversion ratio to obtain the cost per unit of raw protein, (which is c' in Chapter 5).

(4) **Cost of Protein Qualitatively Adjusted**—The conversion ratio just described does not take into account the varying efficiencies of protein by food type. As suggested above, protein efficiency varies primarily because of differences in the number of essential amino acids present. The chemical score (H in Chapter 5) is equal to the quantity of the most limiting amino acid in the food protein expressed as a percentage of the same amino acid present in the amino acid scoring pattern of Table 4-9.

In order therefore, to find the true cost of protein—that is, protein qualitatively adjusted into efficiency units—the costs of raw protein (adjusted for wastage) were multiplied by the reciprocal of the index numbers of Table 4-10. The same process was applied to the aquacultured-species protein weights. This provides the qualitatively adjusted cost of protein which—from the standpoint of effective nutrition—is the relevant cost for our consideration. A sample of these adjusted costs appears in Table 4-11, in which each protein sector is ranked from least-cost to highest.

In this chapter we have defined the competing protein sectors,

Table 4-11. Quality-Adjusted Protein Costs by Type, Tanzania*

Protein Sector	Cost (Dollars per Kilogram)
Fin Fish	$0.72
Coarse Grains	1.21
Rice	1.52
Shellfish	2.10
Wheat	2.15
Poultry	4.27
Beef	4.55
Pork	5.06

*Similar data on all nations were generated by the authors.

identified the costs of the factors of production, developed the initial aquacultured models (and hence the parameters for simulation), and defined and justified the use of qualitative adjustment factors in deriving unit protein costs. The next chapter will detail the model in which these data will provide the basis for our simulation results.

REFERENCES

Burk, M.C. and Ezekiel, M., "Food and Nutrition in Developing Economies," in Southworth and Johnson, eds., *Agricultural Development and Economic Growth*, Ithaca, N.Y.: Cornell University Press, 1967.

FAO Agricultural Commodities Projections for 1975 and 1985, Volume 1, Rome, 1957.

FAO Food Balance Sheets, 1964-66, Food and Agricultural Organization of the United Nations, Rome, 1971.

FAO Production Yearbook, 1972, Food and Agricultural Organization of the United Nations, Rome, 1972.

International Financial Statistics, International Monetary Fund, Washington, D.C., 1970.

Protein Requirements, Report of the Food and Agricultural Organization of the United Nations, Rome, 1957.

Appendix 4-A

Food Price Data for Latin America

CLASS II, REGION B: LATIN AMERICA

If not otherwise noted on authors' data sheets, prices are determined as the following:

Commodity	Type of Price	¢/k.g.
Wheat	Average regional producer price	9.2
Rice	Average regional producer price	14.9
Maize	Average regional producer price	8.1
Potatoes	Average regional producer price	9.7
Sugar	Average regional producer price	12.0
Vegetables	Average regional producer price	9.4
Fruits	Average regional producer price	10.8
Beef and Veal	Average regional producer price	93.5
Sheep and Lamb	Average regional producer price	92.8
Pork	Average regional producer price	77.5
Poultry	Average regional producer price	51.8
Eggs	Average regional producer price	101.3
Fish	Average regional producer price	22.8
Milk	Average regional producer price	13.2
Fats and Oils	Average regional producer price	54.5
Wheat	Average regional producer price	9.2
Wheat	Argentina producer price	4.1
Wheat	Guatemala producer price	12.8
Wheat	Peru producer price	10.6
Rice	Average regional producer price	14.9
Rice	Argentina producer price	3.9
Rice	Costa Rica producer price	28.2

Commodity	Type of Price	¢/k.g.
Rice	Guatemala producer price	14.7
Rice	Peru producer price	12.7
Maize	Average regional producer price	8.1
Maize	Argentina producer price	4.5
Maize	Costa Rica producer price	9.8
Maize	Guatemala producer price	9.5
Maize	Peru producer price	8.5
Sorghum	Average regional producer price	5.8
Sorghum	Argentina producer price	2.9
Sorghum	Costa Rica producer price	5.9
Sorghum	Peru producer price	8.5
Potatoes	Average regional producer price	9.7
Potatoes	Guatemala producer price	13.6
Potatoes	Peru producer price	5.8
Sweet Potatoes	Average regional producer price	7.1
Sugar	Average regional export price (cane sugar)	12.0
Sugar	Caribbean ports	8.2
Sugar	Mexico	15.7
Pulses (dry beans)	Average regional producer price	24.8
	Costa Rica regional producer price	28.2
	Guatemala regional producer price	25.4
	Peru regional producer price	20.7
Dry Peas	Average world producer price	8.7
Lentils	Average regional producer price (assumed equal to Peru price)	14.4
Broad Beans	Average regional producer price (assumed equal to Italy price)	11.5
Chick peas	Average regional producer price (assumed equal to Peru price)	22.5
Seeds and Nuts (ground nuts)	Average regional price (assumed equal to Peru producer price)	20.3
Almonds	Average world producer price	57.5
	U.S. price	71.2
	Italy price	43.8
Coconuts	Philippines export price	39.0
Vegetables	Average world producer price for onions and tomatoes (8.8) (10.0)	9.4
Fruits	Average world producer price for apples, oranges, pears, lemons (9.1) (10.6) (10.2) (13.1)	10.8
Beef and Veal	Average world wholesale price (slaughtered weight)	93.5
Sheep and Goats	Average world wholesale price (slaughtered weight)	92.8

Commodity	Type of Price	¢/k.g.
Pigs	Average world producer price (slaughtered weight)	77.5
Poultry	Average world wholsale price (chickens slaughtered price)	51.8
Eggs	Average regional producer price	101.3
	Costa Rica price	110.1
	Guatemala price	92.6
Fish (fin fish)	Average world producer price	22.8
Fish (shellfish)	Assumed same as fin fish	22.8
Milk	Average regional producer price	13.2
	Costa Rica price	7.2
	Guatemala price	14.0
	Puerto Rico price	18.4
Fats and Oils	Average world wholesale price (see Africa explanation)	54.5

✼ *Appendix 4-B*

Factor Price Data by Country

Country	Major Crop	Income & Budget Share Multiplier	Annual[1] Wage($)	Rental Value of Land	Rental Rate on Capital	Price per Unit of Fertilizer
Algeria	Wheat	99.0	$225.0	$151.21	9.8	This column is a constant for each country at $.08/kg.
Angola	Coarse grains	99.0	40.5	380.38	9.8	
Cameroon	Coarse grains	93.6	158.4	15.17	9.8	
Egypt	Wheat	99.0	157.5	572.27	10.0	
Ethiopia[2]	Coarse grains	56.7	18.0	58.83	9.8	
Ghana[2]	Potatoes & starchy food	99.0	96.3	782.71	10.5	
Guinea[2]	Potatoes & starchy food	84.6	36.0	61.00	9.8	
Ivory Coast[2]	Potatoes & starchy food	99.0	90.0	32.49	9.8	
Kenya[2]	Coarse grains	93.6	29.7	234.58	9.8	
Madagascar[2]	Potatoes & starchy food	84.6	31.5	41.46	9.8	
Malawi[2]	Coarse grains	56.7	28.8	35.70	9.8	
Mali[2]	Coarse grains	84.6	30.6	4.48	9.8	
Morocco	Wheat	99.0	186.3	86.49	8.5	
Mozambique[2]	Potatoes & starchy food	99.0	76.5	195.67	9.8	
Nigeria[2]	Potatoes & starchy food	84.6	63.9	102.43	9.5	
Rhodesia[2]	Coarse grains	99.0	45.0	158.55	9.8	
South Africa	Coarse grains	186.3	189.0	94.78	9.5	
Sudan	Potatoes & starchy food	84.6	32.4	65.97	9.8	
Tanzania[2]	Potatoes & starchy food	84.6	45.9	9.61	9.8	
Tunisia[2]	Wheat	99.0	30.6	30.40	10.0	
Uganda[2]	Potatoes & starchy food	84.6	81.0	51.58	9.8	
Upper Volta[2]	Coarse grains	56.7	22.5	8.64	9.8	
Bolivia	Potatoes & starchy food	93.6	35.1	45.95	14.8	
Brazil	Potatoes & starchy food	168.3	350.1	21.99	25.0	
Chile	Wheat	186.3	453.6	24.13	25.0	

Columbia	Potatoes & starchy food	102.0	207.9	376.28	19.0
Ecuador	Potatoes & starchy food	99.0	207.9	68.72	13.0
Peru	Potatoes & starchy food	168.3	441.9	68.54	14.5
Venezuela	Potatoes & starchy food	285.6	610.2	78.66	10.0
Greece	Wheat	285.6	763.2	233.95	11.5
Hungary	Wheat	285.6	743.4	227.25	10.5
Poland	Potatoes & starchy food	285.6	633.6	170.66	10.5
Portugal	Wheat	186.3	369.0	118.00	8.5
Romania	Wheat	285.6	755.1	8.36	10.5
Spain	Potatoes & starchy food	285.6	584.1	33.29	11.5
Yugoslavia	Wheat	186.3	676.8	270.81	10.5
Chad[2]	Coarse grains	56.7	21.6	8.57	9.8
Congo[2]	Potatoes & starchy food	84.6	123.3	8.57	9.8
Dahomey[2]	Potatoes & starchy food	84.6	60.3	25.98	9.8
Gabon[2]	Potatoes & starchy food	168.3	277.2	3725.62	9.8
Libya	Wheat	417.0	2247.3	373.11	10.0
Niger[2]	Coarse grains	84.6	35.1	9.84	9.8
Sierra Leone[2]	Rice	93.6	24.3	30.77	9.8
Costa Rica	Potatoes & starchy food	168.3	201.6	242.64	10.0
El Salvador	Coarse grains	99.0	202.5	353.12	11.1
Honduras	Coarse grains	99.0	225.9	144.26	14.8
Nicaragua	Coarse grains	102.0	188.1	185.12	11.1
Panama	Potatoes & starchy food	186.3	303.3	425.59	14.8
Cyprus	Wheat	186.3	492.3	74.79	15.4
Israel	Wheat	469.5	1156.5	395.07	13.5
Jordan	Wheat	102.0	288.9	46.24	10.3

Country	Major Crop	Income & Budget Share Multiplier	Annual[1] Wage($)	Rental Value of Land	Rental Rate on Capital	Price per Unit of Fertilizer
Laos[2]	Rice	84.6	36.9	46.97	15.4	
Lebanon	Wheat	168.3	439.2	1294.73	15.4	
Albania	Wheat	168.3	342.0	152.43	10.5	
Iceland	Finfish	417.0	919.8	108899.36	10.3	
Zaire[2]	Potatoes & starchy food	84.6	108.0	68.82	9.8	
Zambia[2]	Coarse grains	102.0	39.6	10.45	9.8	
Cuba	Wheat	168.3	261.0	243.93	14.8	
Dom. Rep.	Potatoes & starchy food	102.0	234.9	133.13	14.8	
Guatemala	Coarse grains	102.0	243.0	207.79	14.8	
Haiti[2]	Potatoes & starchy food	84.6	26.1	442.91	14.8	
Mexico	Coarse grains	186.3	468.9	18.86	9.5	
Afganistan[2]	Wheat	84.6	24.3	16.27	15.4	
Bangladesh[2]	Rice	93.6	27.9	263.35	15.4	
Burma[2]	Rice	84.6	30.6	32.99	15.4	
Mainland China	Rice	84.6	121.5	96.96	15.4	
Taiwan	Rice	102.0	136.8	2.57	17.5	
Hongkong	Rice	186.3	290.7	5918.90	15.4	
India	Rice	84.6	105.3	54.51	10.4	
Indonesia	Rice	84.6	150.3	109.33	15.4	
Iran	Wheat	102.0	251.1	73.30	13.0	
Iraq	Wheat	102.0	200.7	41.61	15.4	
Khmer[2]	Rice	93.6	37.8	178.13	15.4	
Korea Demo.	Coarse grains	99.0	180.0	384.46	15.4	
Korea Rep.	Rice	99.0	222.3	322.04	24.0	
Malaysia	Rice	102.0	191.7	45.30	15.4	
Nepal[2]	Rice	84.6	21.6	286.95	15.4	

Pakistan²	Rice	93.6	27.9	394.45	15.4
Philippines	Rice	99.0	171.9	141.76	15.0
Saudi Arabia	Coarse grains	186.3	350.1	2232.23	15.4
Sri Lanka	Rice	84.6	160.2	1937.65	15.4
Syria	Wheat	99.0	293.4	18.14	10.0
Thailand	Rice	93.6	105.3	40.63	10.0
Turkey	Wheat	168.3	546.3	72.97	14.0
N. Vietnam	Rice	84.6	117.0	93.89	15.4
S. Vietnam	Rice	93.6	156.6	115.80	15.4
Yemen²	Coarse grains	56.7	198.0	8.57	15.4
Argentina	Potatoes & starchy food	285.6	540.0	75.76	14.8
Paraguay	Potatoes & starchy food	99.0	74.7	192.59	14.8
Uraguay	Wheat	186.3	305.1	102.49	14.8

1. Wages are derived from the assumptions listed in the test.
2. An opportunity cost of one cent per kilogram was used as a proxy for major crop price.

Prices of Other Production Inputs

Price Per Unit of Feed*
(macrobrachium)

Algeria	$10.37	Honduras	$10.12
Angola	4.25	Nicaragua	10.12
Cameroon	8.50	Panama	10.12
Egypt	9.13	Cyprus	10.25
Ethiopia	8.50	Israel	10.25
Ghana	4.25	Jordan	10.25
Guinea	4.25	Laos	10.25
Ivory Coast	4.25	Lebanon	10.25
Kenya	4.88	Albania	11.62
Madagascar	4.25	Iceland	11.62
Malawi	6.75	Zaire	4.25
Mali	8.50	Zambia	4.25
Monocco	10.37	Cuba	10.12
Mozambique	4.25	Dominican Rep.	10.12
Nigeria	8.50	Guatemala	11.88
Rhodesia	4.25	Haiti	10.12
South Africa	4.25	Mexico	11.25
Sudan	14.12	Afganistan	17.37
Tanzania	5.75	Bangladesh	10.25
Tunisia	10.37	Burma	10.25
Uganda	3.50	Mainland China	10.25
Upper Volta	8.50	Taiwan	10.25
Bolivia	10.12	Hong Kong	10.25
Brazil	276.25	India	11.12
Chile	10.12	Indonesia	10.25

Columbia	10.12	Iran	10.25
Ecuador	10.12	Iraq	10.25
Peru	11.37	Khmer	10.25
Venezuela	10.12	Korea Demo.	10.25
Greece	11.62	Korea Rep.	11.37
Hungary	11.37	Malaysia	10.25
Poland	11.37	Nepal	10.25
Portugal	11.62	Pakistan	10.25
Romania	11.62	Philippines	10.25
Spain	11.37	Saudi Arabia	14.12
Yugoslavia	7.25	Sri Lanka	10.25
Chad	8.50	Syria	9.25
Congo	4.25	Thailand	10.25
Dahomey	4.25	Turkey	9.50
Gabon	4.25	N. Vietnam	10.25
Libya	9.25	S. Vietnam	10.25
Niger	5.75	Yeman	14.12
Sierra Leone	5.75	Argentina	5.63
Costa Rica	12.25	Paraguay	10.12
El Salvador	10.12	Uruguay	10.12

*Price per unit of feed is constant for all species except macrobrachium. The price for all countries and other species is $.04/kg.

Price per Unit of Fingerlings*

	Mullet	*Walking Catfish*	*Yellowtail*	*Eel*	*Carp*	*Tilapia*
Algeria	$.01	$.03	$.01	$.03	$.02	$.01
Angola	.00	.00	.00	.07	.00	.00
Cameroon	.00	.02	.00	.03	.01	.01
Egypt	.00	.02	.00	.06	.01	.01
Ethiopia	.00	.00	.00	.02	.00	.00
Ghana	.00	.01	.00	.11	.01	.00
Guinea	.00	.00	.00	.02	.00	.00
Ivory Co	.00	.01	.00	.03	.01	.00
Kenya	.00	.00	.00	.00	.00	.00
Madagasc	.00	.00	.00	.11	.00	.00
Malawi	.00	.00	.00	.01	.00	.00
Mali	.00	.00	.00	.07	.00	.00
Morocco	.01	.02	.00	.09	.02	.01
Mozambiq	.00	.01	.00	.04	.01	.00
Nigeria	.00	.01	.00	.04	.01	.00
Rhodesia	.00	.01	.00	.09	.00	.00
S Africa	.01	.02	.00	.12	.02	.01
Sudan	.00	.00	.00	.15	.00	.00
Tanzani	.00	.01	.00	.15	.00	.00
Tunisia	.00	.00	.00	.13	.00	.00
Uganda	.00	.01	.00	.07	.01	.00
Upper Vo	.00	.00	.00	.15	.00	.00
Zaire	.00	.01	.00	.12	.01	.00
Zambia	.00	.00	.00	.14	.00	.00
Cuba	.01	.03	.01	.00	.02	.01
Dom Rep	.01	.03	.01	.03	.02	.01
Guatemal	.01	.03	.01	.01	.02	.01
Haiti	.00	.00	.00	.06	.00	.00
Mexico	.01	.05	.01	.46	.04	.02
Afghanis	.00	.00	.00	.01	.00	.00
Banglade	.00	.00	.00	.00	.00	.00
Burma	.00	.00	.00	.04	.00	.00
China Ma	.00	.01	.00	.04	.01	.00
China Ta	.01	.03	.01	.05	.02	.01
Hong Kon	.01	.03	.01	.04	.02	.01
India	.00	.01	.00	.06	.01	.00
Indonesa	.00	.02	.00	.10	.01	.01
Iran	.01	.03	.01	.23	.02	.01
Iraq	.01	.02	.00	.06	.02	.01
Khmer	.00	.00	.00	.01	.00	.00
Korea DP	.01	.02	.00	.09	.02	.01
Korea R	.01	.03	.01	.07	.02	.01
Malaysia	.01	.02	.00	.19	.02	.01
Nepal	.00	.00	.00	.02	.00	.00
Pakistan	.00	.00	.00	.06	.00	.00

	Mullet	Walking Catfish	Yellowtail	Eel	Carp	Tilapia
Philippi	.01	.02	.00	.05	.01	.01
Sua Arab	.01	.04	.01	.01	.03	.01
Sri Lank	.00	.02	.00	.03	.01	.01
Syr Ar R	.01	.03	.01	.03	.02	.01
Thailand	.00	.01	.00	.00	.01	.00
Turkey	.02	.06	.01	.02	.05	.02
V Nam DR	.00	.01	.00	.01	.01	.00
V Nam R	.00	.02	.00	.02	.01	.01
Yem Ar R	.00	.00	.00	.01	.00	.00
Argentin	.02	.06	.01	.01	.05	.02
Bolivia	.00	.00	.00	.01	.00	.00
Brazil	.01	.04	.01	.01	.03	.01
Chili	.01	.05	.01	.04	.04	.02
Columbia	.01	.02	.00	.02	.02	.01
Ecuador	.01	.02	.00	.01	.02	.01
Peru	.01	.05	.01	.01	.04	.02
Venezuel	.02	.07	.01	.04	.05	.02
Greece	.02	.09	.02	.01	.06	.03
Hungary	.02	.08	.02	.01	.06	.03
Poland	.02	.07	.02	.01	.05	.02
Portugal	.01	.04	.01	.02	.03	.01
Romania	.02	.09	.02	.00	.06	.03
Spain	.02	.07	.01	.02	.05	.02
Yugoslav	.02	.08	.02	.01	.06	.02
Chad	.00	.00	.00	.05	.00	.00
Congo	.00	.01	.00	.05	.01	.00
Dahomey	.00	.01	.00	.05	.01	.00
Gabon	.01	.03	.01	.01	.02	.01
Lib Ar R	.07	.25	.05	.10	.19	.08
Niger	.00	.00	.00	.00	.00	.00
Si Leone	.00	.00	.00	.01	.00	.00
Cos Rica	.01	.02	.00	.01	.02	.01
El Salva	.01	.02	.00	.02	.02	.01
Honduras	.01	.03	.01	.06	.02	.01
Nicaragu	.01	.02	.00	.08	.02	.01
Panama	.01	.03	.01	.02	.03	.01
Cyprus	.01	.06	.01	.03	.04	.02
Israel	.03	.13	.03	.05	.10	.04
Jordon	.01	.03	.01	.04	.02	.01
Laos	.00	.00	.00	.01	.00	.00
Lebanon	.01	.05	.01	.04	.04	.02
Albania	.01	.04	.01	.05	.03	.01
Iceland	.03	.10	.02	.04	.08	.03
Paraguay	.00	.01	.00	.00	.01	.00
Uruguay	.01	.03	.01	.01		.01

*Price per unit of feed is constant for all species except macrobrachium. The price for all countries and species is $.04/kg. A reported price of zero in this table means that the price is less than one cent per unit.

Chapter Five

Aquaculture Technology Transfer: A Simulation Model

BIOECONOMIC MODEL

It is fairly obvious that many aquaculture enterprises are extremely complex operations sometimes involving the production of fry or fingerlings and/or the mixture of two products in joint production. In this chapter we specify a general bioeconomic model that may be applicable to the "typical" aquaculture enterprise. The model itself is more or less a systems approach to evaluating the impact of differing factor prices and environmental variables on the transfer of aquaculture technology from one area of the world to another. The following *terms* will be used:

k = Developing countries (DC's); $(k_1, k_2 \ldots k_{90})$

n = aquaculture species of farms within each k $(n_1, n_2 \ldots n_{14})$

v = protein producing sectors within each k other than aquaculture; $(v_1, v_2 \ldots v_{16})$

C_{nk} = total cost of production for a "typical" enterprise of n'th species in k'th DC

Q_{nk} = total output (gross landed weight in kilograms) for the "typical" enterprise of n'th species in k'th DC

K_{nk} = capital input of n'th species in k'th DC

L_{nk} = labor input of n'th species in k'th DC

D_{nk} = land input of n'th species in k'th DC

Z_{nk} = fertilizer input of n'th species in k'th DC

F_{nk} = feed input of the n'th species in the k'th DC

S_{nk} = stocking of (fingerlings, fry) input of the n'th species in the k'th DC

I_{nk} = other economic inputs of n'th species in k'th DC[1]

where p_K, p_L, p_D, p_Z, p_F, p_S, and p_I are input prices for *capital, labor, land fertilizer, feed, stocking* and other *inputs* respectively for the aquaculture farm.

As explained in Chapter 2, one of the most critical barriers in developing a viable aquaculture enterprise is the physical environment. Thus, we must consider differences in the physical environment from country to country and how these differences influence each type of species considered for aquaculture transfer. E_1, E_2, $E_3 \ldots E_m$ are the necessary environmental variables running through m in number such as water temperature, PH, salinity and oxygen requirements as discussed in Chapter 2. These environmental variables apply to the n'th species in the k'th DC or $(E_1)_{nk}$; $(E_2)_{nk}$ $\ldots (E_d)_{nk}$. We may write the general relation between total output for the n'th species in the k'th DC and inputs discussed above as

$$Q_{nk} = f(K, L, D, Z, F, S, I, E_1, E_2 \ldots E_m) \qquad (5.1)$$

where the non-market environmental inputs $(E_1, E_2 \ldots E_m)$ enter the production process (or function) and ultimately the cost of production.

In this simplified relation in (5.1) no economies of scale are assumed. That is, large aquaculture farms produce at about the same average cost of production as smaller farms, holding other factors constant. The literature and data in this area are presently insufficient to test the hypothesis that unit costs change with scale. Also, in our simulation model we will not be dealing with changes in the size of the farm, but rather with the potentiality for transferring a *given size* aquaculture farm from one area of the world to another. Admittedly, certain areas may be more suitable for a larger or smaller aquaculture enterprise; however, this is beyond the scope of our analysis and data base.

We will also make another simplifying assumption of no factor substitution if relative factor prices change from country to country. As with economies of scale, no data are readily available to test this assumption since it is doubtful that much factor substitution is

---

1. Economic inputs are defined as those factors of production which have an external positive price. The firm either pays for their use or some opportunity cost.

known for such small and simple enterprises with which we are dealing. For a discussion of this assumption, see Arrow, Chenery, Solow and Minhas (1961). To still further simplify the analysis, we will view $E_1, E_2 \ldots E_d$ as *production function shifters.*[2]

The value of E is determined in the following way:

$0 < E < 1 =$ environmental variable decreases production (percent less than 100) in country from which aquaculture technology has been transferred;

$E = 1 \quad =$ environmental variable does not change production from which aquaculture technology has been transferred;

$E > 1 \quad =$ environmental variable increases production over country from which aquacultural technology has been transferred.

Therefore (5.1) may be respecified as

$$Q_{nk} = f(K, L, D, F, S, I)E_1, E_2 \ldots E_m, \quad (5.2)$$

and the total cost of output unadjusted for environmental changes for the "typical" aquaculture enterprise in any LDC is

$$C_{nk} = [p_K K + p_L L + p_D D + p_Z Z + p_F F + p_S S + p_I I]. \quad (5.3)$$

Equation (5.3) specifies the economic dimension of technology transfer, but must be combined with environmental considerations.

As indicated above, there are a number of critical environmental variables that may influence the level of production after the technology has been transferred. The research team's biological and environmental experts *combined the influence* of $E_1, E_2 \ldots E_m$ for each country into *one* value for E (see Appendix 5-A for actual values by country). Therefore, we have the following simplified concept for species n in DC k:

$$\left[\frac{C}{EQ}\right]_{nk} = \frac{\text{Economic Costs}}{\text{Environmentally Adjusted Output}}. \quad (5.4)$$

Equation (5.4) merely specifies our estimate of the *unit cost of*

2. As an example, if we evaluated rainfall to be above that normally occurring in the country from which the technology is being transferred we increased (decreased) the productivity or output (Q_{nk}) by a certain percent depending upon the environmental analysis.

production for the aquaculture enterprise adjusted for different
(1) factor prices (reported in Appendixes 4-B and 4-C), and (2) environmental conditions compared to the country of origin. Of course,
where $E = 0$, no transfer is possible.

ADJUSTMENTS TO GROSS COST PER UNIT OF OUTPUT

Equation (5.4) will only yield cost per unit of gross output. That is,
output is *unadjusted* for (1) percent of fish and/or plant yielding
edible protein (M) which gives protein fish weight, and (2) the
quality of protein or the amino acid pattern of the species. As
defined in Chapter 4, the quality of protein by a chemical score (H)
will have a maximum of 100 percent. Therefore, the qualitatively-
adjusted cost is derivable with H. (The values of M and H for each of
the sample species are presented in Table 5.1). Thus, we have the
following:

$$\left[\frac{C}{EQ}\right]_{nk} \cdot M^{-1} = \left[\frac{C}{EQ}\right]'_{nk} = \text{cost per unit of protein} \quad (5.5)$$
$$\text{fish weight}$$

$$\left[\frac{C}{EQ}\right]'_{nk} \cdot H^{-1} = \left[\frac{C}{EQ}\right]''_{nk} = \text{cost per unit of quali-} \quad (5.6)$$
$$\text{tatively adjusted protein.}$$

To simplify the notation the values defined in Equations (5.5) and
(5.6) will be specified as c'_{nk} and c''_{nk} respectively. Furthermore, let
us specify the v'th producing protein sector (non-aquaculture) within
each k as c'_{kv} and c''_{kv} which correspond to our Equations (5.5) and
(5.6). For each k, we shall have two vectors:

Cost per unit of protein weight	Cost per unit of qualitatively adjusted protein
$(c'_{nk})_1$	$(c''_{nk})_1$
$(c'_{nk})_2$	$(c''_{nk})_2$
.	.
.	.
.	.
$(c'_{nk})_{14}$	$(c''_{nk})_{14}$
$(c'_{nv})_1$	$(c''_{nv})_1$
$(c'_{nv})_2$	$(c''_{nv})_2$
.	.
.	.
.	.
$(c'_{nv})_{16}$	$(c''_{nv})_{16}$

Table 5-1. Protein Content and Chemical Scores for Sample Species

Species	Share of Gross Edible	×	Kilograms of Protein per kg. of Edible Portion (M)	=	Adjusted kg. of Protein per kg. of Gross Weight	Chemical Score (H)	Quality-adjusted Protein per kg.
1. Common Carp	.53		.103		.055	100	.055
2. Catfish (Channel)	.59		.182		.107	100	.107
3. Tilapia*	.74		.160		.118	100	.118
4. True Eel	.77		.168		.129	89	.115
5. Rainbow Trout	.70		.206		.144	80	.115
6. Mullet	.52		.207		.108	100	.108
7. Milk fish*	.69		.206		.142	100	.142
8. Yellowtail	.77		.217		.167	100	.167
9. Penaeus Shrimp	.51		.176		.090	95	.086
10. Oyster	.18		.109		.020	77	.015
11. Mussels (baby clam)	.36		.076		.027	100	.027
12. Walking Catfish*	.59		.182		.107	100	.107
13. Macrobrachium*	.51		.176		.090	95	.086
14. Blue-Green Algae**	.66		.025		.017	74	.013

Source: *Food Composition Table for Use in East Asia* (1972).
*Assumed to be same as for related species.
**Assumed to be the same as asparagus.

In simplified language, there are three measures of average unit cost of production: (1) gross; (2) flesh weight yield and (3) quality of protein yield. The latter two measures give the user a better idea of the cost of protein. Also, the two vectors above indicate that within any k, we can compare the cost of protein from our 14 aquaculture enterprises with 16 other protein producing sectors which may be presently producing protein in the k'th country. Hence, one can make comparisons to see which kind of farming enterprise delivers protein at the lowest cost (e.g., a carp farm vs. a pig farm).

DEMAND AND OTHER FACTORS

The bioeconomic model discussed above omits several critical factors. Up to this point we have ignored consumer acceptance of the transferred products. To do a complete analysis of acceptability one would have to conduct a marketing study for each DC. Obviously, this is beyond the scope of this study. However, we did make some preliminary social analysis based on secondary data for each DC as to probable acceptance of "fish in general." Given the nature of our analysis, there is no way to quantify the level of potential demand, except in areas where fish is completely prohibited on religious or other social grounds. Where acceptability is ruled out, of course, consumer substitution of aquaculture for other protein sources becomes a moot concept. The question how does aquaculture compare with other protein producing sectors essentially implies that protein (qualitatively adjusted) is *homogeneous.* Yet adjustments need to be made where our consumer analysis shows fish or certain species of fish to be unacceptable (i.e., zero demand cross elasticity). Without information on demand cross elasticities, it is difficult to make a quantitative estimate of potential consumption of aquacultured products due to substitution, but an approximation method will be attempted in Chapter 6. Of course, areas of complementarity may be found where substitution is not necessary. Also, we have said nothing about a nation's ability to be a technology receiver. Obviously different countries are at different levels of technological advancement. We need therefore to consider the relation of the technological state to transferability.

All this suggests that once costs of production are estimated, we have three stages of analysis. (1) *Substitutability*: Are aquaculture costs sufficiently low to substitute a species for some alternative protein source? (2) *Demand*: Even if aquaculture costs are relatively low, will the species experience consumer acceptability? (3) *Technology*: Even if aquaculture costs are low, a species is acceptable, does the nation have a technology state favorable for the species transfer?

THE ESTIMATION OF POTENTIAL PRODUCTION, EMPLOYMENT AND FOREIGN EXCHANGE FROM AQUACULTURE

For the k'th developing nation, we estimated (1) available coastline or coastal zone in hectares that could be used for saltwater aquaculture farms and (2) available inland area available for aquaculture for those species that also can be cultivated in fresh water. These estimates appear in Chapter 7. In each case, there may be present use of the coast or inland areas for other economic activities and especially food production. For example, we could assume that the entire coastal area would be used for penaeus shrimp production. This would be an obvious overstatement. This information led us to some *very tentative estimates* of hectares used by species in each DC that appear in Chapter 7. From this, we derived potential *production, employment and foreign exchange* per hectare for each developing nation.

SUMMARY

In this chapter we have developed the basic bioeconomic model for simulation purposes. As we have shown, such costs must take into account the protein quantity and quality of the producing sectors. The parameter estimates from our model enterprises and the factor price data presented earlier are combined in this cost simulation model in order to make intersector and intercountry comparisons. Nonetheless, once the potential for substitution is assessed we have noted that other considerations are important. We therefore need to build into our evaluation the potential for consumer acceptability, and the potential for new technology acceptance. These stages of the systems analysis are detailed in the next chapter.

REFERENCES

Arrow, K., B. Chenery, B. Minhas and R. Solow, "Capital-Labor Substitution and Economic Efficiency," *Rev. Econ. Stat.* (August 1961).

Food Composition Table for Use in East Asia, UNFAO and USHEW, Rome and Washington, 1972.

**Environmental Acceptability
(*E*) for Fourteen Aquaculture
Enterprises by Country**

Countries	Indian Carp	Channel Catfish	Tilapia	True Eel	Rainbow Trout	Mullet	Milk-fish	Fish Yellow-tail	Panaeus Shrimp	Oysters	Mussels	Walking Catfish	Macrobro-chium	Blue-Green Algae
1. Afghanistan	0	0	0	0	1.0	0	0	0	0	0	0	0	0	0
2. Albania	1.0	1.0	1.0	.8	1.0	1.0	.2	1.0	1.0	1.0	1.0	.5	1.0	1.0
3. Algeria	0	.8	1.0	0	.2	1.2	.2	0	1.0	1.0	1.0	1.0	.8	.8
4. Angola	.5	1.0	1.0	.2	0	.9	.8	.8	.8	.8	.8	1.0	.8	.8
5. Argentina	.5	.3	.2	.3	1.5	0	0	0	1.0	.3	.3	.3	1.0	1.0
6. Bangladesh	1.0	1.0	1.0	1.0	0	1.0	.8	.8	.8	.8	.8	1.0	1.0	.8
7. Bolivia	.2	.7	1.0	1.0	1.0	0	0	0	0	1.0	0	.7	.8	0
8. Brazil	1.0	1.0	1.0	1.0	0	1.0	1.0	1.0	1.0	1.0	1.0	1.0	1.0	1.0
9. Burma	1.0	1.0	1.0	1.0	.5	1.0	1.0	1.0	.7	.7	.7	1.0	1.0	.3
10. Cameroon	1.0	1.0	1.0	1.0	0	1.0	.8	1.0	1.0	1.0	1.0	1.0	1.0	1.0
11. Chad	1.0	1.0	1.0	0	0	0	0	0	0	0	0	1.0	1.0	0
12. Chile	0	.2	.2	.3	1.2	1.0	0	1.0	0.5	1.2	1.2	0	1.0	1.0
13. China (Mainland)	.6	.8	.8	1.0	1.0	1.0	1.0	1.0	1.0	1.0	1.0	.7	1.0	1.0
14. China (Taiwan)	1.0	1.0	1.0	1.0	0	1.0	1.0	1.0	1.0	1.0	1.0	1.0	1.0	1.0
15. Columbia	1.0	1.0	1.0	.8	1.0	1.0	1.0	1.0	1.2	1.0	1.0	1.0	1.0	1.0
16. Congo	1.0	1.0	1.0	1.0	0	1.0	1.0	1.0	1.0	1.0	1.0	1.0	1.0	1.0
17. Costa Rica	1.0	1.0	1.0	.8	.5	1.0	.3	1.0	1.2	1.0	1.0	1.0	1.0	1.0
18. Cuba	1.0	1.0	1.0	.8	0	1.0	.3	1.0	1.0	1.0	1.0	1.0	1.0	1.0
19. Cyprus	1.0	1.0	1.0	.5	0	1.0	.2	1.0	1.0	1.0	1.0	1.0	1.0	1.0
20. Dahomey	1.0	1.0	1.0	.8	0	1.2	1.0	1.2	1.2	1.0	1.0	1.0	1.0	1.0
21. Dominican Repub.	1.0	1.0	1.0	.8	0	1.0	.3	1.0	1.2	1.0	1.0	1.0	1.0	1.0
22. Ecuador	1.0	1.0	1.0	.8	1.0	1.0	1.0	1.0	1.0	.7	.7	1.0	1.0	1.0
23. Egypt	1.0	1.0	1.0	1.0	0	1.0	0	1.0	1.0	1.0	1.0	1.0	1.0	1.0
24. El Salvador	1.0	1.0	1.0	1.0	.5	1.0	1.0	1.0	1.0	1.0	1.0	1.0	1.0	1.0
25. Ethiopia	1.0	1.0	1.0	1.0	.6	1.0	.2	1.0	.5	.5	.5	1.0	1.0	1.0
26. Gabon	1.0	1.0	1.0	1.0	0	1.0	.8	1.0	1.0	1.0	1.0	1.0	1.0	1.0
27. Ghana	1.0	1.0	1.0	1.0	0	1.2	.8	1.2	1.2	1.0	1.0	1.0	1.0	1.0
28. Greece	1.0	1.0	1.0	.8	1.0	1.0	.2	1.0	1.0	1.0	1.0	.5	1.0	1.0

#	Country																
29.	Guatemala	1.0	1.0	1.0	1.0	1.0	1.0	.5	1.0	1.0	1.0	1.2	1.0	1.0	1.0	1.0	1.0
30.	Guinea	1.0	1.0	1.0	1.0	1.0	1.0	0	1.0	1.0	1.0	1.0	1.0	1.0	1.0	1.0	1.0
31.	Haiti	1.0	1.0	.8	1.0	1.0	1.0	.5	.3	1.0	1.0	1.0	1.0	1.0	1.0	1.0	1.0
32.	Honduras	1.0	1.0	1.0	1.0	1.0	1.0	.5	.3	1.0	1.0	.8	1.0	1.0	1.0	1.0	1.0
33.	Hong Kong	1.0	1.0	1.0	1.0	1.0	1.0	1.0	1.0	1.0	1.0	.8	1.0	1.0	1.0	1.0	1.0
34.	Hungary	0	0	0	0	0	0	0	0	0	0	0	0	0	0	0	0
35.	Iceland	0	0	0	0	0	0	1.2	0	1.0	1.0	.5	.8	.5	0	0	0
36.	India	1.0	1.0	1.0	1.0	1.0	1.0	0	1.0	1.0	1.0	1.0	1.0	1.0	1.0	1.0	1.0
37.	Indonesia	1.0	1.0	1.0	1.0	1.0	1.0	0	1.0	1.0	1.0	1.0	1.0	1.0	1.0	1.0	1.0
38.	Iran	1.0	1.0	1.0	1.0	1.0	1.0	.5	.5	1.0	1.0	1.0	.3	1.0	1.0	1.0	1.0
39.	Iraq	1.0	1.0	1.0	1.0	1.0	1.0	0	0	1.0	1.0	1.0	.3	1.0	1.0	1.0	1.0
40.	Israel	1.0	1.0	1.0	1.0	1.0	1.0	.8	.2	1.0	1.0	1.0	1.0	.5	1.0	1.0	1.0
41.	Ivory Coast	1.0	1.0	1.0	1.0	1.0	1.0	0	1.0	1.2	1.0	1.2	1.0	1.0	1.0	1.0	1.0
42.	Jordan	1.0	1.0	1.0	1.0	1.0	1.0	1.0	.2	1.0	1.0	1.0	1.0	.5	1.0	1.0	1.0
43.	Kenya	1.0	1.0	1.0	1.0	1.0	1.0	1.0	1.0	1.0	1.0	1.0	1.0	1.0	1.0	1.0	1.0
44.	Khmer	1.0	1.0	1.0	1.0	1.0	1.0	0	1.0	1.0	1.0	1.0	1.0	1.0	1.0	1.0	1.0
45.	Korea (DPR)	.5	0	1.0	1.0	1.0	1.0	0	.5	1.0	1.0	1.0	1.0	.5	1.0	1.0	1.0
46.	Korea Republic	0	.2	1.0	1.0	1.0	1.0	0	0	1.0	1.0	1.0	1.0	0	1.0	1.0	1.0
47.	Laos	1.0	1.0	1.0	1.0	1.0	1.0	0	1.0	1.0	1.0	1.0	1.0	1.0	1.0	1.0	1.0
48.	Lebanon	0	.8	.5	.5	1.0	1.0	0	.2	1.2	1.0	1.0	1.0	1.0	1.0	1.0	1.0
49.	Libyan Arab Repub.	1.0	1.0	1.0	1.0	1.0	1.0	0	.2	1.0	1.0	1.0	1.0	1.0	.8	.8	.8
50.	Madagascar	1.0	1.0	1.0	1.0	1.0	1.0	0	0	1.0	1.0	1.0	1.0	1.0	1.0	1.0	1.0
51.	Malawi	1.0	1.0	1.0	1.0	1.0	1.0	0	0	1.0	1.0	1.0	1.0	1.0	1.0	1.0	1.0
52.	Malaysia	1.0	1.0	1.0	1.0	1.0	1.0	0	1.0	1.0	1.0	1.0	1.0	1.0	1.0	1.0	1.0
53.	Mali	1.0	1.0	1.0	1.0	1.0	1.0	0	1.0	1.0	1.0	1.0	1.0	1.0	1.0	1.0	1.0
54.	Mexico	1.0	1.0	1.0	1.0	1.0	1.0	1.0	1.0	1.0	1.0	1.0	1.0	1.0	1.0	1.0	1.0
55.	Morocco	.3	.5	1.0	1.0	1.0	1.0	.5	.5	1.0	1.0	1.0	1.0	.5	1.0	1.0	1.0
56.	Mozambique	1.0	1.0	1.0	1.0	1.0	1.0	0	0	1.0	1.0	1.0	1.0	1.0	1.0	1.0	1.0
57.	Nepal	.8	.5	0	.8	1.0	1.0	1.0	1.0	1.0	1.0	1.0	1.0	1.0	.3	1.0	1.0
58.	Nicaragua	1.0	1.0	1.0	1.0	1.0	1.0	.5	.5	1.0	1.0	1.2	1.0	1.0	1.0	1.0	1.0
59.	Niger	1.0	1.0	1.0	1.0	1.0	1.0	1.0	0	1.0	1.0	0	1.0	1.0	1.0	1.0	1.0
60.	Nigeria	1.0	1.0	1.0	1.0	1.0	1.0	0	1.0	1.0	1.0	1.0	1.0	1.0	1.0	1.0	1.0

Countries	Indian Carp	Channel Catfish	Tilapia	True Eel	Rainbow Trout	Mullet	Milk-fish	Yellow-tail	Panaeus Shrimp	Oysters	Mussels	Walking Catfish	Macrobrochium	Blue-Green Algae
								Fish						
61. Pakistan	1.0	1.0	1.0	1.0	0	1.0	1.0	1.0	1.0	1.0	1.0	1.0	1.0	1.0
62. Panama	1.0	.5	1.0	1.0	.5	1.0	1.0	1.0	1.0	1.0	1.0	1.0	1.0	1.0
63. Paraguay	.5	.5	1.0	0		0		0		1.0	1.0	1.0	1.0	1.0
64. Peru	1.0	1.0	1.0	.8	1.0	1.0	.5	1.0	1.0	.7	.7	1.0	1.0	1.0
65. Philippines	1.0	1.0	1.0	1.0	0	1.0	1.0	1.0	1.0	1.0	1.0	1.0	1.0	1.0
66. Poland	0	0	0	.8	1.0	.5	0	0	0	.5	.5	0	0	0
67. Portugal	.2	.5	0	.8	.6	1.0	.8	1.0	1.0	1.0	1.0	0	1.0	1.0
68. Rhodesia	1.0	1.0	1.0	0	.8	0	0	0	0	0	0	1.0	.8	0
69. Romania	0	.5	0	.8	1.0	1.0	0	0	0	0	0	0	0	.5
70. Saudi Arabia	0	.2	.8	.5	0	1.0	0	.2	.2	.2	.2		1.0	.5
71. Sierra Leone	1.0	1.0	1.0	1.0	0	1.0	1.0	1.0	1.0	1.0	1.0		1.0	1.0
72. South Africa	.8	1.0	1.0	1.0		1.0	1.0	1.0	1.0	1.0	1.0	.5	1.0	1.0
73. Spain	.2	.5	0	.5	.9	1.0	.8	1.0	1.0	1.0	1.0	1.0	1.0	1.0
74. Sri Lanka	1.0	1.0	1.0	1.0	1.0	1.0	1.0	1.0	1.0	1.0	1.0	1.0	1.0	1.0
75. Sudan	1.0	1.0	1.0	.5	0	1.0	1.0	1.0	0	0	0	1.0	1.0	1.0
76. Syrian Arab Repub.	0	.5	.8	.5	0	1.0	0	.2	.5	.5	.5	1.0	1.0	.5
77. Tanzania	1.0	1.0	1.0	1.0	0	1.0	1.0	1.0	1.0	1.0	1.0	1.0	1.0	1.0
78. Thailand	1.0	1.0	1.0	1.0	0	1.0	1.0	1.0	1.0	1.0	1.0	1.0	1.0	1.0
79. Tunisia	1.0	.8	1.0	0	.2	1.0	.2	0	1.0	.8	1.0	1.0	1.0	.8
80. Turkey	0	.5	0	1.0	1.0	1.0	0	1.0	.8	.8	.8	0	1.0	1.0
81. Uganda	1.0	1.0	1.0	0	0	0	0	0	0	0	0	1.0	1.0	0
82. Uruguay	.8	.5	1.0	1.0	0	1.0	1.0	1.0	1.0	1.0	1.0	.8	1.0	1.0
83. Upper Volta	1.0	1.0	1.0	0	0	0	0	0	1.0	0	0	1.0	1.0	0
84. Venezuela	1.0	1.0	1.0	.8	.8		1.0	1.0	1.2	1.0	1.0	1.0	1.0	1.0
85. Viet Nam (Dem Rep)	1.0	1.0	1.0	1.0	0	1.0	1.0	1.0	1.0	1.0	1.0	1.0	1.0	1.0
86. Vietnam (Rep of)	1.0	1.0	1.0	1.0	0	1.0	1.0	1.0	1.0	1.0	1.0	1.0	1.0	1.0
87. Yemen (Arab Repub)	0	.2	.8	.5	1.0	1.0	0	.2	.2	.2	.2	0	1.0	.5
88. Yugoslavia	0	.5	0	.8	1.0	1.0	.2	1.0	1.0	1.0	1.0	1.0	1.0	1.0
89. Zaire	1.0	1.0	1.0	0	0	0	0	0	0	0	0	1.0	1.0	0
90. Zambia	1.0	1.0	1.0	0	0	0	0	0	0	0	0	1.0	1.0	0

Sources: Analysis by Randy Martin, Department of Biological Sciences, Florida State University; and Mr. Glassen, Research Associate, Center for Resource and Environmental Analysis, Florida State University. They used various biological works, topology maps, and advice from various experts to arrive at an approximation of environmental acceptability.

 *Chapter Six*

The Transfer of Aquaculture Technology: Simulation Results

INTRODUCTION

In this chapter, we will combine the simulation results on protein costs (c'_{nk}) and adjusted protein costs (c''_{nk}) with other variables to derive the information necessary for making species recommendations for potential technology receivers. Even though on an environmental basis (see Chapter 5) the country is suitable for technology transfer, there are other relevant considerations such as demand or consumer acceptability and the ability of the local populace to utilize the new technology. This chapter therefore is devoted to (1) the theory of decision making and (2) the feasibility results.

A SYSTEMS FRAMEWORK: THE DECISION MATRICES

The following discussion describes the development of three decision matrices that are used in selecting species and technology receivers.

1. A *substitution matrix* is first constructed from computations of c' and c'' for each country (k). This will include *all* protein producing sectors (i.e., both hypothetical aquaculture (n) and existing agriculture (v)). Sectors are ranked from the lowest cost of protein in turn, adjusted and unadjusted, to highest for each k. A sample ranking for the kth country might be as follows: $(c'_{n_2 k})_1$; $(c'_{n_5 k})_2$; $(c'_{v_8 k})_3$; $\ldots (c'_{m_{10} k})_{23}$; and in turn: $(c''_{n_3 k})_1$; $(c''_{v_5 k})_2$; $(c''_{n_4 k})_3$; $\ldots (c''_{v_9})_{23}$. In each case, the subscripted number denotes

the lowest (1st) to highest cost ranking possible (23th). The reader will recall that we began with 16 non-aquaculture protein-producing sectors. But seven of these sectors were found to be insignificant producers so that (counting the 14 selected species) the maximum number of protein producing sectors per country is 23. This, of course, does not mean that all 23 sectors will be producing protein in any particular nation.

We next code the protein sectors in the following way. All ranked producing sectors are divided into quintiles with the lowest cost one-fifth coded "A," the lowest one-fifth coded "E," and all coded "A" → "E." Where species costs were not significantly different at the margin between codes, we used independent judgments in moving them from a "higher-cost" to a "lower-cost" code. Table 6-1 provides an illustration of these rankings and costs for Honduras. In Honduras there is significant production in nineteen protein-producing sectors. Among the lowest-cost protein sectors are two of our selected species, mullet ($.07 per kilogram) and Rainbow Trout ($1.70 per kilogram). Both receive "A" codes. Among other protein producing sectors, Walking Catfish is coded C while Yellowtail,

Table 6-1. Adjusted Costs by Protein Sector, Honduras

Sector	U.S. Dollars per Kilogram	Substitution Code
Mullet	$ .07	
Wheat	1.28	
Rice	1.59	A
Rainbow Trout	1.70	
Coarse Grains	2.53	
Oysters	2.75	
Poultry	3.06	B
Beef	3.51	
Channel Catfish	6.09	
Tilapia	7.39	
Pork	8.00	C
Walking Catfish	8.45	
Yellowtail	13.48	
Milkfish	16.98	
Penaeus Shrimp	24.93	D
Macrobracium	25.45	
Mussels	33.03	
Shell Fish (wild stock)	40.00	E
True Eel	43.82	

Panaeus Shrimp, Macrobrachium and Milkfish are coded D. Mussels and True Eel are high cost sectors and appear in the highest-cost quintile (E).

The above procedure has several advantages. First, we know from the quality of the data (see Chapters 2 and 3) that our results are likely to have wide variances. By ranking an aquaculture enterprise in the upper 20 percent of the low cost sectors we are allowing for variances and also reducing the evaluation of the substitution possibilities to five grades, A → E.

The above information is used to construct the substitution matrix. The matrix elements are the *code scores* given to each species by country. Any inference exclusively from this matrix requires the assumption that not only protein but the products themselves are homogeneous across sectors (i.e., infinite demand cross-elasticity between products). This is obviously an unrealistic assumption since demand is a significant factor and cross elasticities of demand are finite. An illustrative substitution matrix is shown below.

In this illustrative example, Indian Carp in Algeria is in the least cost category of protein while Tilapia is in the highest cost category. However, in Angola the situation is reversed. Two substitution

Aquaculture Sectors

Country	Indian Carp	Tilapia	Oysters	 14
1. Algeria	A	E	C	
2. Angola	E	A	B	
.				
.				
.				(etc.)
.				
.				
90.				

Figure 6-1. The Aquaculture Substitution Matrix

matrices are relevant in the study, one for c' and the other for c''. The *un*adjusted protein costs of protein are given in Appendix 6-A for the current protein-producing sectors and the simulated costs for the lowest cost sectors (A) appear in Appendix 6-B.

2. A *demand* or *consumer acceptability matrix* is next developed. The substitution matrix shows the ordinal degree of substitution that *may* be possible for aquacultured products based solely upon low c's and c''s for aquaculture species compared to the other sectors for a particular country (k). It is important to account for demand differences—especially those based upon tastes—before the final selection of species with a high potential for technology transfer.

An extensive sociological analysis of consumer acceptability was conducted to estimate consumer acceptability. Certain generalizations can be drawn from this analysis. In the Hamito-Semitic cultures of North Africa, there is a general dietary prohibition against non-scale fish of any sort. This prohibition is associated with Judeo-Islamic religious principles and may be extended to all areas where Islam is found.

A general observation also may be made about Sub-Saharan Africa. Black Africa has various tribal taboos against fish; however, they do not carry strong religious prohibitions. In some instances, these taboos are directed only to women or children. The reasoning is that delicacies are reserved for adults or men. In addition, Africans generally prefer *larger* fish as pointed out by Bardach (1972).

Wherever there is a specific religious-vegetarian cultural syndrome, an obvious barrier to fish consumption exists. Within this syndrome Hindu and Buddhists constitute a significant proportion of world population. It is a strong factor in the Indian subcontinent or Indo-Asia and there are substantial portions of the urban population of East Africa that are of Indian origin.

In Latin America, the low incidence of fish in the diet is essentially attributable to the fact that fish has never been part of their tradition even in coastal areas, despite fairly ample fishery resources. No specific prohibitions are evident. The Indian subculture of Latin America shows a pattern of fish consumption that is sporadic and based largely on pragmatism (i.e., only animal protein available).

Of great import, two of the major barriers to fish consumption are *inadequate processing technology* and *transportation facilities* which greatly inhibit regional distribution.

Again, an index is constructed. The descriptions of consumer demand conditions are as follows:

C = The species listed is *currently cultured* in the country.

G = The species has a *good* chance of acceptance as a transferred species. That is, the species is acceptable by sociocultural criteria of the major part of the population. The species is not prohibited on religious or other grounds and the country in question is presently consuming similar or related species.

P = This species is not expressly prohibited, but it is considered that the transfer would be risky or *poor*. In this category, we have various taboos that are less than a specific religious prohibition. In addition, there are generally depreciative attitudes toward fish as a human food.

N = The species is expressly and specifically prohibited as a matter of religious faith.

A linear numerical score attached to each of these codes extends from highest probability for acceptance to zero probability for acceptance. An abbreviation of the description of the demand condition, the associated letter code and a related numerical score are presented in Table 6-2. An index number, therefore, for the nth species and the kth country can be assigned.

The index numbers by country and species comprise one demand acceptability matrix which is illustrated in Figure 6-2.

3. The third decision matrix is a *technology acceptability* (i.e., capacity to accept or adapt to new technology) *matrix*. The index of technological acceptability is a composite of (1) the country ranking of investment shares of GNP for 1970; (2) an assignation of a score to each country based upon the Harbison-Myers (1964) index of human resources and (3) selected sociological indicators from Adelman and Morris (1967). All these appear in Appendix 6-C: *Technological Acceptance Indicators*.

Table 6-2. Demand Acceptability Codes and Indices

Demand Condition	Code	Assigned Numerical Index
Already cultured	C	3
Satisfactory potential acceptance	G	2
Poor potential acceptance	P	1
Unacceptable	N	0

Source: Sociological Analysis of Consumer Acceptability by Gary Enoch.

Country	Indian carp	tilapia	oysters	Aquaculture Sectors 14
1. Algeria	3	1	2	
2. Angola	0	1	0	
.				
.				
.				
.				
.				
90.				(etc.)

Figure 6-2. The Demand Acceptability Matrix (A Sample)

The matrix consists of the elements derived from each of these sources. Investment as a share of GNP is widely considered as an index of the level of technological sophistication of a nation. The Harbison-Myers Index is an education development measure that emphasizes higher education. (The index combines secondary enrollment ratios and higher education ratios, the latter weighted by a factor of five.) This measure of "schooling capacity" is properly viewed as a proxy for technical skills and knowledge of the population.

The balance of the technological elements is derived from the pathbreaking contribution of Irma Adelman and Cynthia Morris. Their method of classification—similar to that of the Harbison-Myers Index—assigns letter classifications similar to classroom grades to each country for that particular attribute. Such a classification allows for a qualitative judgment where quantification is not especially relevant. The measures selected as matrix columns are (a) the extent of literacy, (b) the level of economic modernization, (c) the commitment to economic growth, (d) the degree of administrative efficiency, (e) the extent of mass communication and (f) the degree of national integration.

We will briefly describe each of these measures. For the extent of literacy, a first approximation is derived from data on the percent of literate adults. But where such data are weak or known to be misleading a qualitative adjustment can be made. The level of economic modernization is based upon the techniques used in agriculture. Specifically, it is a measure of the improvements in the organization of agriculture as the relative importance of traditional and modern agricultural sectors changed between 1950 and 1960. The letter grades assigned to the extent of leadership commitment to economic development are derived from judgments of experts about governments' behavior during the early 1960s. In the developing countries it is presumed that private sector activity alone will not rapidly advance agricultural technology and that strong national leadership is required. Likewise, the degree of administrative efficiency is a measure of the ability of the governments to implement their development programs.

The extent of mass communication is derived from a composite index based on daily newspaper circulation and the number of radio receivers. This is an indicator of the rate at which technical knowledge can be dispersed. And, finally, the degree of national integration is a rough measure of the social cohesion of the society and its ability to exert common effort toward goal achievements.

Where data are not available the matrix cells are empty in Appendix 6-C. In the few cases where insufficient information is available, the country is assigned a score equal to that of the neighboring country under the assumption that a neighbor most nearly resembles the country's technical possibilities.

If sufficient detail were available about the technology requirements of each species, a technology matrix would be available for each species. This is not the case and the existing matrix is "summed" horizontally into a vector of technology indicators. The vector consists of country scores that are assigned as follows:

I = Most Potential Acceptance to Technology
II = Above Average Potential Acceptance to Technology
III = Average Potential Acceptance to Technology
IV = Below Average Potential Acceptance to Technology
V = Least Potential Acceptance to Technology

These scores are illustrated below in a schema of the technology acceptability matrix (Figure 6-3). In this illustration different species

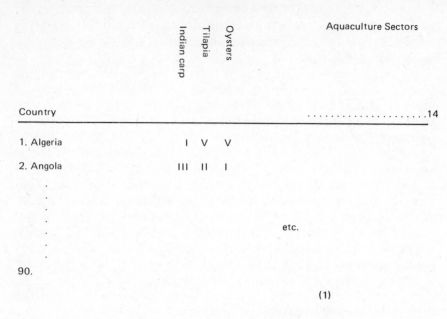

Figure 6-3. The Technology Acceptability Matrix

have variant rankings in each country. As indicated above, because of limited information, we retain only a vector of rankings in which only one rank is assigned to each nation. In Figure 6-3, Algeria is seen to be very receptive to the technology of Indian Carp culture and very low in receptivity to Tilapia and Oysters culturing. In our final rankings both Algeria and Angola had actual rankings of III.

4. The *SDT Matrix* contains the elements of the *substitution, demand* and *technological acceptability* matrices combined. It provides the *overall* score for each species by country and is the final decision matrix. For example, an A,3,I score would indicate the highest probability for successful technology transfer. The SDT Matrix is illustrated in Figure 6-4. Note that—because *T* is a vector—only one *Technology Acceptability Score* is given for each nation.

The above procedure provides a systematic method to evaluate the potential of species for transfer to augment protein supplies. At this juncture, we assume that aquaculture is substituted for other crops on the basis of cost, demand and technological considerations. However, complementarity may also exist where there is no competition between land or sea area. It is beyond the scope of this study to consider land and sea availability in every one of our 90 countries in precise detail. Nonetheless, the FSU Center for Resources and

| Country | Indian Carp | | | Tilapia | | Aquaculture Sectors |
	S	D	T	S	D	14
1. Algeria	*	2	III	C	2	
2. Angola	E	2	III	C	2	
.						
.						
.					(etc.)	
.						
.						
90.						

*Indicates no entry because the environmental variable (E) = 0.

Figure 6-4. The SDT Matrix of Combined Scores

Environmental Analysis did give us a rough indication of land and/or sea coast available for aquaculture. This availability and its implications are discussed in Chapter 7.

In some cases where aquaculture scores well in the SDT matrix, there may be a conflict between the use of the end product as an *export* or for *domestic* consumption. Conceivably, a species with a high-cost substitution matrix score could be selected for production and marketing on the international market. If species n in the k'th country were given a substitution score of E, it is not necessarily ruled out for aquaculture. Say country k can culture the species at a cost of $x per kilogram while the world price is $y per kilogram and $x < $y. There may be a possibility that the revenue derived from foreign exchange from species n will be sufficient so that the country can displace wheat production, for example, with the aquaculture operation and import more wheat with the foreign exchange earnings. This type of possibility also will be evaluated in Chapter 7.

COMBINED SPECIES SCORE: DECISION RULES FOR THE FEASIBILITY OF TRANSFER

Table 6-3 shows the code score for each element of the substitution matrix (S). Continuing our earlier example, under "S" in the table is

Table 6-3. Country Rankings by Indices of Substitution, Consumer Acceptability and Implementation Capability for Fourteen Aquaculture Species[1]

Country	Indian Carp S	D	T	Channel Catfish S	D	Tilapia S	D	True Eel S	D	Rainbow Trout S	D	Mullet S	D	Milk Fish S	D	Yellowtail S	D	Penaeus Shrimp S	D	Oysters S	D	Mussels S	D	Walking Catfish S	D	Macrobrakium S	D	Blue-Green Algae S	D
1. Algeria	*	2	III	C	0	C	2	*	0	B	0	A	2	D	1	*	1	D	0	B	0	D	0	C	0	D	0	E	1
2. Angola	E	2	III	C	1	C	2	E	1	*	1	A	1	C	1	D	1	D	1	B	1	C	1	C	1	D	1	E	1
3. Cameroon	E	2	III	B	3	A	3	E	1	*	1	A	2	B	1	C	1	C	2	A	1	D	1	C	2	D	2	E	1
4. Egypt	E	2	II	C	0	C	3	D	0	*	0	A	2	*	1	C	1	D	0	B	0	D	0	B	0	D	0	E	1
5. Ethiopia	E	1	IV	B	1	A	1	E	1	A	1	A	1	D	1	C	1	C	3	A	1	C	1	C	1	D	2	E	1
6. Ghana	E	2	II	C	1	D	2	D	1	*	1	A	1	C	1	C	1	D	1	B	1	C	1	C	1	D	2	E	1
7. Guinea	E	2	IV	B	1	B	2	E	1	*	1	A	1	B	1	D	1	C	1	A	1	C	1	C	1	D	1	E	1
8. Ivory Coast	E	2	III	B	1	B	3	E	1	*	1	A	1	B	1	D	1	C	1	A	1	D	1	C	1	D	1	E	1
9. Kenya	E	2	III	C	1	C	2	D	1	*	1	A	1	C	1	D	1	C	1	B	1	C	1	C	1	D	1	E	1
10. Madagascar	E	2	III	C	1	B	3	E	1	*	1	A	1	B	1	D	1	C	1	A	1	C	1	D	1	D	1	E	1
11. Malawi	E	3	III	C	3	B	3	*	1	*	1	*	1	*	1	*	1	*	1	*	1	*	1	D	2	E	1	E	*
12. Mali	E	2	III	B	1	A	2	*	1	*	1	*	1	*	1	*	1	*	1	*	1	*	1	E	1	E	1	E	*
13. Morocco	*	2	III	C	0	C	2	E	1	A	0	A	2	B	1	C	1	E	0	B	0	D	0	D	0	D	0	E	1
14. Mozambique	E	3	II	C	1	C	3	D	1	*	1	A	1	B	1	C	1	D	1	A	1	C	1	C	1	D	1	E	1
15. Nigeria	E	3	III	C	1	B	2	E	1	*	1	A	3	B	1	D	1	D	3	A	1	C	1	C	1	D	2	E	1
16. Rhodesia	E	2	II	D	1	D	3	*	1	A	1	A	1	*	1	*	1	*	1	*	1	E	1	D	1	E	1	E	*
17. S. Africa	E	2	I	C	1	C	3	E	1	A	1	A	1	B	1	D	1	D	1	B	1	E	1	C	1	D	1	E	1
18. Sudan	E	3	III	B	1	B	3	E	1	*	1	A	1	*	1	D	1	D	1	A	1	D	1	C	1	D	1	E	1
19. Tanzania	E	3	III	C	1	A	3	E	1	*	1	A	1	B	1	D	1	B	1	A	1	D	1	D	1	D	1	E	1
20. Tunisia	*	2	II	C	0	A	2	B	0	B	0	A	3	D	1	*	1	B	0	A	0	B	0	D	0	D	0	E	1
21. Uganda	E	2	III	C	1	B	3	*	1	*	1	*	1	*	1	*	1	*	1	*	1	D	1	D	1	E	1	E	*
22. Upper Volta	E	2	V	B	1	B	3	*	1	*	1	*	1	*	1	*	1	*	1	*	1	*	1	D	1	E	1	E	*

23. Zaire	E	2	III	C	1	B	*	A	*	A	1	*	A	1	*	A	*	B	*	D	3	B	1	C
24. Zambia	E	2	III	B	1	B	*	A	*	A	1	*	A	1	*	B	1	B	*	E	3	B	1	B
25. Cuba	E	1	II	B	1	C	2	A	2	C	2	2	D	2	2	A	2	C	2	E	2	C	1	B
26. Dom Rep	E	3	III	B	1	B	2	A	2	D	2	3	D	2	2	A	2	D	2	E	3	B	1	B
27. Guatemala	E	3	III	B	1	C	2	A	2	D	3	3	D	2	2	A	2	D	3	D	1	C	1	B
28. Haiti	E	1	V	C	1	B	2	A	1	B	2	2	C	2	2	A	1	D	2	E	2	C	1	C
29. Mexico	E	1	IV	*	1	D	2	A	2	D	2	2	D	2	2	A	2	D	2	D	1	C	1	*
30. Afghanistan	*	2	IV	C	2	B	*	A	3	B	2	3	B	2	2	A	3	E	1	E	1	C	2	C
31. Bangladesh	E	3	IV	C	2	B	2	A	3	C	2	3	C	2	2	A	3	B	2	D	2	B	2	B
32. Burma	E	2	II	B	2	D	2	A	3	B	2	3	D	3	3	A	3	B	2	E	3	B	3	B
33. China Ma	E	3	II	B	3	D	3	A	3	B	3	3	D	3	3	A	3	B	3	D	3	B	2	C
34. China Ta	E	3	I	C	2	D	2	A	3	B	2	2	D	2	2	A	3	B	2	D	3	E	2	C
35. Hong Kon	*	3	I	D	2	B	2	A	3	C	3	3	E	2	2	A	3	E	2	E	2	E	0	D
36. India	E	3	I	B	2	E	2	A	3	D	2	3	B	3	3	A	3	B	2	D	3	B	3	C
37. Indonesa	E	3	II	B	2	D	2	A	3	E	2	2	C	2	2	A	3	C	2	E	2	B	2	C
38. Iran	E	2	II	B	0	B	2	A	3	D	2	2	D	2	2	A	3	*	2	E	2	B	0	C
39. Iraq	E	3	III	B	0	E	2	A	3	E	2	2	E	2	2	A	3	B	2	D	2	B	0	E
40. Khmer	E	3	III	B	3	*	2	A	2	B	2	2	B	2	2	A	2	C	2	D	2	B	0	C
41. Korea Dr	E	2	III	*	2	B	2	A	2	D	2	2	C	2	2	A	3	B	2	E	2	*	0	E
42. Korea R	*	3	II	*	3	D	2	A	2	D	2	2	D	2	2	A	2	*	2	D	2	*	0	C
43. Malaysia	E	3	III	B	3	B	2	A	3	D	2	2	B	2	2	A	3	B	2	D	3	B	3	B
44. Nepal	E	2	III	E	2	B	0	A	3	D	2	2	D	2	2	A	3	*	0	E	2	E	3	E
45. Pakistan	E	3	II	C	3	C	0	A	3	E	2	2	C	2	2	A	0	C	2	D	2	C	3	C
46. Philippines	E	3	I	C	3	E	0	A	2	D	2	2	D	2	2	A	2	B	2	*	2	C	0	B
47. Sau Arab	*	2	III	E	0	C	0	A	0	*	0	0	D	0	0	A	0	D	0	D	0	E	0	*
48. Sri Lank	E	0	II	D	0	E	0	A	0	C	0	0	C	0	0	A	2	*	0	E	3	C	0	C
49. Syr Ar R	*	2	II	C	0	C	0	A	0	D	0	0	D	0	0	A	3	B	0	E	3	B	1	*
50. Thailand	E	3	I	B	3	B	3	A	3	D	3	3	D	3	3	A	2	*	3	E	2	B	0	*
51. Turkey	*	2	I	D	0	*	1	A	0	E	0	0	D	1	1	A	2	*	1	E	0	*	0	*

Table 6-3. (cont.)

	Indian Carp		Channel Cat-fish	Tilapia	True Eel	Rainbow Trout	Mullet	Milk Fish	Yellowtail	Penaeus Shrimp	Oysters	Mussels	Walking Cat-fish	Macro-brakium	Blue-Green Algae
52. V Nam Dr	E 2	III	B 2	B 2	E 2	* 2	A 2	B 2	C 2	D 2	A 2	D 2	C 2	D 2	E 2
53. V Nam R	E 3	III	B 3	B 2	E 2	* 2	A 2	B 3	C 2	D 2	A 2	D 2	C 2	D 2	E 2
54. Yem Ar R	* 2	V	D 0	A 2	E 0	* 0	A 2	* 1	E 1	* 0	A 0	D 0	* 0	0 0	D 1
55. Argentin	* 2	I	D 1	C 2	E 1	A 1	A 2	* 2	* 2	D 2	D 2	* 2	D 1	D 1	* 1
56. Bolivia	* 1	III	D 1	C 2	* 1	A 2	* 2	* 3	* 2	D 2	* 2	* 2	E 1	E 2	* 1
57. Brazil	E 1	II	C 3	C 2	D 1	* 2	A 2	B 2	C 2	D 2	B 2	E 2	C 1	D 2	E 1
58. Chili	* 2	I	D 1	D 2	E 1	* 2	A 2	* 2	C 2	D 2	B 2	D 2	* 1	D 2	E 1
59. Columbia	E 1	I	C 1	D 2	E 1	A 2	A 2	C 2	D 2	D 2	B 2	D 2	D 1	D 2	E 1
60. Ecuador	E 1	II	C 1	C 1	E 1	A 2	A 2	B 2	D 2	D 2	B 2	E 2	C 1	D 2	E 1
61. Peru	E 1	II	B 1	B 3	E 1	A 2	A 2	C 2	D 2	D 2	B 2	E 2	C 1	D 2	* 1
62. Venezuela	E 1	I	C 1	C 1	E 1	A 2	A 2	B 2	D 2	D 2	B 3	E 2	C 1	D 2	* 2
63. Greece	* 1	*	C 1	D 1	* 1	* 2	* 3	D 2	* 2	E 2	C 2	E 2	D 1	* 2	* 1
64. Hungary	* 3	*	* 3	* 1	E 2	A 2	A 2	* 2	* 2	D 2	* 2	* 2	* 2	* 2	* 2
65. Poland	* 3	*	* 2	* 1	E 2	A 3	A 2	* 2	* 2	E 2	D 2	E 2	* 2	* 2	* 2
66. Portugal	* 1	I	D 1	* 2	E 2	A 2	A 2	C 2	D 2	E 2	B 3	* 2	* 1	E 1	* 2
67. Romania	* 3	I	D 2	* 2	E 2	A 2	A 2	* 2	* 2	E 2	* 2	E 2	* 2	* 2	* 2
68. Spain	* 1	I	C 1	* 2	* 1	A 2	A 2	C 2	D 2	E 2	C 3	* 3	* 1	E 1	* 1
69. Yugoslavia	* 3	I	D 2	* 2	E 1	A 2	A 3	D 2	E 2	E 2	C 3	* 2	* 2	E 2	* 1
70. Chad	E 2	IV	* 1	A 3	* 1	* 1	* 1	* 1	D 1	* 2	* 1	* 1	D 1	E 1	* 1
71. Congo	E 2	IV	B 3	A 3	E 1	* 1	A 1	B 1	D 1	* 1	A 1	D 1	D 2	E 1	B 1
72. Dahomey	E 2	IV	C 1	B 3	D 1	* 1	A 1	B 1	D 1	C 3	A 1	C 1	C 1	D 1	E 1
73. Gabon	* 2	III	D 1	* 3	D 1	* 1	A 1	E 1	D 1	D 1	D 1	E 1	C 1	D 1	* 1
74. Lib Ar R	* 2	III	C 0	D 2	* 0	* 0	A 2	D 2	E 1	E 0	D 0	E 0	D 0	0 0	* 1
75. Niger	* 2	IV	C 1	A 2	* 1	* 1	* 1	* 1	* 1	D 1	* 1	* 1	D 1	E 1	* 1

76. Si Leone	E 2	III	C 1	B 2	E 1	* 1	A 1	B 1	D 1	B 1	A 1	B 1	C 1	D 1	0 1
77. Cos Rica	E 1	I	B 1	C 3	E 1	A 2	A 2	C 2	C 2	D 2	A 2	C 2	D 2	E 1
78. El Salva	* 1	II	D 3	D 3	E 1	A 2	A 2	D 2	D 1	E 3	B 2	C 2	D 2	* 1
79. Honduras	* 1	III	C 3	C 1	E 1	A 2	A 2	D 2	D 2	D 3	B 2	C 2	D 2	* 1
80. Nicaragu	E 1	III	B 1	C 1	E 2	A 2	A 2	B 2	D 2	D 2	B 2	C 2	D 1	E 1
81. Panama	* 1	II	C 1	D 3	E 1	A 2	A 2	C 2	D 2	E 3	B 2	C 2	E 1	* 2
82. Cyprus	E 1	I	B 1	B 2	E 1	* 2	A 3	D 2	D 2	E 3	B 2	O 2	D 2	* 2
83. Israel	* 3	I	C 0	C 3	D 0	* 0	A 3	D 3	C 1	O 0	C 0	O 0	D 0	* 1
84. Jordon	E 2	II	B 0	B 3	E 0	* 2	A 2	C 2	C 1	B 0	B 0	C 0	D 0	E 1
85. Laos	* 2	III	C 2	B 2	* 2	* 2	A 2	* 2	* 2	* 2	* 2	D 2	* 2	* 2
86. Lebanon	* 2	I	C 0	D 2	D 0	* 0	A 2	D 2	C 1	C 0	D 0	C 0	D 0	* 2
87. Albania	E 1	II	B 1	B 1	E 2	A 2	A 2	D 2	C 1	E 2	B 2	C 2	D 2	E 2
88. Iceland	* 1	I	* 1	* 1	* 1	* 2	C 2	B 2	* 1	* 2	* 2	* 2	* 2	* 1
89. Paraguay	* 2	III	D 1	C 2	* 1	* 1	A 2	D 2	* 2	* 2	B 2	D 1	E 2	* 1
90. Uruguay	* 2	I	D 1	C 2	E 1	* 1	A 2	B 2	D 2	B 2	B 2	C 1	D 2	E 1

1. T = Technology Acceptability Index, which is the same for each species; S = Substitution code; D = Demand Acceptability Index.
*No entry because E = 0.
Source: See Appendixes 6.B and 6.C.

the letter A for Mullet and Rainbow Trout in Honduras (79.) which indicates a high level of substitution. The raw data from which the substitution matrix is constructed are presented in Appendix 6-B: *Adjusted Cost of Protein per Kilogram by Sector by Country*. The fourteen adquaculture sectors were ranked along with the present protein producing sector for each country. The "other" protein sectors were reduced to the nine sectors upon which nations depend for the bulk of their protein supply. Also, in the final empirical results, we consider only c''—the quality-adjusted cost (or price) of protein per kilogram—since this is the most relevant protein measure for dietary reasons. Hence, c', although defined in the decision-making sections above, will be omitted in order to directly address the major problem of dietary protein.

Table 6-3 also shows the indices (3, 2, 1, and 0) for the demand acceptability matrix (D). Under "D," 2's are given to Mullet and Rainbow Trout in Honduras, meaning a good potential for acceptance by consumers. As indicated earlier, the degree of consumer acceptability ranges from 0 to 3. The highest value, 3, signifies that some culturing of the species already is ongoing. As discussed earlier, data on species cultured by country are difficult to obtain. Hence, the 3 might merely indicate initial introduction of the species. Nonetheless, it does connote a high evaluation for potential consumer acceptability.

The elements of the technology acceptability matrix were derived from Appendix 6-C: *Technological Acceptance Indicators*. Table 6-3 shows the results. For example, Honduras has a technological acceptability score (or index) of III which is shown under "T." This means that Honduras has an average potential acceptance to technology for *all* species. This rating is not specific to any one species.

The best combined or SDT score achievable by any species is an A,3,I. (1) This means that the indigenous *level of technology of the kth country is favorable to receive a new food production type.* (2) Upon the transfer of species n to k country the species *would be relatively cheap to produce;* (3) The country has already underway some aquaculturing which indicates *an entrepreneurial expectation of consumer acceptance.* Obviously these all are extremely important considerations in the successful transfer of technology.

In the example of Honduras the combined SDT score for Mullet is (A,2,III) and for Rainbow Trout (A,2,III). This means that, for both species, there are average technological conditions for the potential receipt of the new type of production. Once adopted by transfer, the costs of production would be very low relative to production costs in other protein producing sectors. Once produced, there is a satisfactory potential for acceptability by consumers. Thus an overall decision criterion is required for developing a "recommended" list of countries and species.

To select species for transfer, we developed the following criteria:

1. *Highest Potential*
   (a) A, 3, I or II or III
   (b) A, 2, I or II or III
   (c) B, 3, I or II or III
   (d) B, 2, I or II or III
2. *Average Potential*
   (a) C, 3, I or II or III
   (b) C, 2, I or II or III

The main difference between *highest* and *average* "potential" for transfer feasibility is the relaxation of the substitution requirement wherein the recommended aquacultured commodity is allowed into the middle of the distribution of protein costs (c''). The reader, of course, can establish his own criteria from Table 6-3 as a decision tool. For example, the reader could base a decision purely on costs and demand acceptability and assume that the requisite technological environment could be "created." Let us now review the results of the selection process.

THE RESULTS OF THE DECISION ANALYSIS

Tables 6-4 and 6-5 show the countries that would qualify as having either excellent or average potential for culturing the species shown in the row at the top. Returning to our example with Honduras, under Mullet and Rainbow Trout we find the "highest" potential for aquaculture technology transfer. Their SDT scores (A, 2, III) place these species in the "highest" category.

In our analysis six species have a high probability for successful substitution (if land is available and not having complementary uses) for presently produced crops. The following could be transferred and yield an increase in protein.

1. Channel catfish
2. Rainbow Trout
3. Mullet
4. Milkfish
5. Oysters
6. Tilapia

Table 6-4.   Countries with Highest Potential for Aquaculture Technology Transfer

Channel Catfish	Rainbow Trout	Mullet		Milkfish	Mussels	Oysters	Tilapia
Cameroon*	Rumania	Algeria	Vietnam DR	Guatemala	Khmer*	Cuba	Cameroon
Burma	Spain	Cameroon	Vietnam R	Mexico*		Dom. Repub.	Ivory Coast
P R China*	Iceland	Egypt	Argentina	Burma		Guatemala	Madagascar
India*	PR China	Morocco	Brazil	PR China*		Mexico	Malawi
Indonesia*	Taiwan*	Nigeria*	Chile	Taiwan*		Burma	Mali
Khmer*	Yugoslavia	Tunisia*	Columbia	India*		China PR	Algeria
Malaysia*	Korea DR	Cuba	Ecuador	Indonesia*		Taiwan	Sudan
Thailand*	Korea R	Dom. Repub.	Paraguay	Khmer*		India	Tanzania
Vietnam DR*	Nepal	Guatemala	Venezuela	Malaysia		Indonesia	Tunisia
Vietnam R*	Bolivia	Mexico	Greece*	Philippines*		Khmer	Uganda
	Columbia	Burma*	Poland	Thailand*		Korea DR	Zaire
	Nicaragua	PR China*	Romania	Vietnam DR		Korea R	Zambia
	Ecuador	Taiwan*	Spain	Vietnam R*		Malaysia	Dom. Rep.
	Honduras	Hong Kong*	Yugoslavia	Brazil*		Philippines	Burma
	Peru	India*	Lib.Arab. Rep	Ecuador		Thailand	China PR
	Panama	Indonesia*	Costa Rica	Venezuela		V. Nam DR	Taiwan
	Greece	Iran	El Salvadore	Nicaragua		V. Nam R	India
	Albania	Iraq	Honduras	Uruguay		Brazil	Indonesia
	Hungary	Khmer*	Nicaragua			Chile	Iran
	Poland*	Korea DR	Panama			Columbia	Iraq
		Korea R	Cyprus*			Ecuador	Khmer
		Malaysia	Israel*			Peru	Malaysia
		Pakistan*	Jordan*			Venezuela	Thailand
		Philippines*	Lebanon			Portugal	V. Nam DR
		Saudi Arab.	Albania			Costa Rica	V. Nam R

Syr. Arab Re.	Uruguay	El Salvador	Peru
Thailand*	Portugal	Honduras	Sierra Leone
Turkey		Nicaragua	Cyprus
		Panama	Jordan
		Cyprus	Laos
		Albania	
		Uruguay	

Note: The following species do not have the highest or average technology transfer using the criteria discussed in the text: Indian carp, true eel, yellowtail, walking catfish, macrobrachium, blue-green algae.
*Presently cultured to varying extent.
Source: Table 6-3.

Table 6-5.   Countries with Average Potential for Aquaculture Technology Transfer

Channel Catfish	Tilapia	Mullet	Milkfish	Yellowtail	Penaeus Shrimp	Mussels	Walking Catfish	Macro- brachium	Oysters
Malawi*	India	Iceland	Cuba	Egypt	Cameroon	Burma	Cameroon	Guatemala	Greece
Taiwan*	Congo*		Dom. Rep.	Cuba	Guatemala	India	Burma	Korea R	Spain
Pakistan*	Algeria		Korea DR	Dom Re.	Burma		PR China		Yugoslavia
Philippines*	Angola		Pakistan	Guatemala	PR China		Taiwan*		
Brazil*	Egypt		Columbia	PR China	India*		Hong Kong		
Honduras*	Kenya		Peru	India	Thailand*		India		
Lacs	Morocco		Portugal	Korea R	Dahomey*		Indonesia		
	Mozambique		Spain	Korea DR			Khmer		
	S. Africa		Costa Rica	Pakistan			Malaysia		
	Cuba		El Salvadore	Philippines			Thailand*		
	Guatemala		Panama	Vietnam R			V. Nam Dr		
	Mexico			Vietnam DR			V. Nam R		
	Pakistan			Brazil			Brazil		
	Philippines			Chile			El Salvadore		
	Syr. Arab. R.			Costa Rica			Honduras		
	Argentina			Albania					
	Bolivia								
	Brazil								
	Costa Rica								
	Israel								
	Paraguay								
	Uruguay								

Note: The following species do not have the average technology transfer using the criteria discussed in the text: Indian Carp, True Eel, Rainbow Trout, Blue-Green Algae.

*Presently cultured to varying extent.

Source: Table 6-3.

To this list can be added the following if we relax our criteria to an average potential.

5. Yellowtail
6. Penaeus Shrimp
7. Walking Catfish

Surprisingly, Indian carp showed little potential for transfer. In India, this type of carp is considered a luxury good; therefore, our data probably reflect its high cost of production. Although the selection of Indian carp is based upon the criteria established in Chapter 2, the unavailability of data is a constraint.

All researchers prior to this study indicate that Tilapia is widely cultured because of its simple technology. This is confirmed by our analysis since Tilapia is transferable with an average or higher potential to over 50 countries. Mullet has an even higher potential for transfer to a similar number of countries.

We should briefly explain the poor showing for some species. Like the Indian carp enterprise, True Eel is a luxury item on the world market and its high price makes it non-competitive with other low cost (c'') protein items within each country. This, of course, does not rule out True Eel for foreign trade, which will be discussed below. Yellowtail and Penaeus Shrimp also command high world prices although Penaeus Shrimp and Yellowtail could be used in many countries for local or domestic consumption.

Oysters and mussels show little potential for transfer primarily because of a low edible-protein conversion ratio (see Chapter 5).

Macrobrachium cost and earnings data were from Hawaii, and some obtained under experimental conditions. Such culturing is at present very high cost. Also, prawns command a high world price. Such factors added to the costs of production estimates. Finally, Blue-Green Algae provides little protein yield and this perhaps rules it out of consideration as a bountiful source of protein. Also, it commands a high world price; therefore, it should be considered for export. The reasons for the low probability for transferring these species are summarized in Table 6-6.

THE EXPORT POTENTIAL FOR AQUACULTURE SPECIES

The export potential for a given species is of necessity related to the world price for it. But only those species that have a substantial commercial market have a "world price." Such markets approximate perfect competition and hence can be assumed "the same every-

Table 6-6.   Characteristics of High-Cost Species

Species	Reason for Low Transferability Probability
Indian Carp	Luxury item
True Eel	Luxury item
Mussels	Low flesh weight yield
Macrobrachium	Luxury item
Blue-Green Algae	Low protein conversion factor

where." Nonetheless no world price is available on Tilapia, and the reader is cautioned that it is very difficult to talk about a "world price" for such species as Oysters, Rainbow Trout, Mullet, Mussels, Walking Catfish and Yellowtail. In any case Table 6-7 offers world price estimates for the species.

In compiling our initial country-specie list for trade potential, the following assumptions are made. (1) The species' simulated domestic production cost is below the estimated world price (Table 6-7). (2) Production for trade requires a higher level of technological

Table 6-7.   World Prices Received by Fishermen for Various Species, 1971*

Species	Dollars (U.S.) per Kilogram
1. Indian Carp	$ .230
2. Walking Catfish	.130
3. Tilapia	NA
4. Penaeus Shrimp	1.170
5. Oysters	.157
6. True Eel	1.750
7. Milkfish	.850
8. Macrobrachium (Prawns)	1.360
9. Channel Catfish	.300
10. Yellowtail	1.250
11. Rainbow Trout	.380
12. Mullet	.260
13. Mussels	.094
14. Blue-Green Algae	.560

*All species expressed in round weight.
Source: FAO Yearbook on Fisheries Statistics, 1971.

capacity than simply the domestic production of the species. (3) Production for export does not compete with production for local alternative protein sources.

From these assumptions Table 6-8 gives the relative prices of species by country where (1) the cost of species production is below the world price and (2) the nation has a Technological Acceptance Score above III. That is, only nations that exhibit the "most potential acceptance to technology" are in the list for export potential. From these data it is clear that five species—Mullet, Milkfish, Oysters, Rainbow Trout, and Penaeus Shrimp—have export potential. Tilapia is, of course, excluded.

Even if the nation can produce a species at a price below the estimated world price, there is no assurance that such a nation will have the motivation and infrastructure to market the species successfully. Moreover, the relative prices tell us very little about land availability and hence the size of productive capacity. Therefore Table 6-8 still does not consider the *level* of potential production based upon land availability. This consideration is brought to the fore in Chapter 7.

To the extent that the species is currently produced in a nation or otherwise experiences domestic consumer acceptability, the production for export would compete with those resources, especially land, that otherwise would be devoted to domestic production. Yet some nations clearly do not have competing domestic consumption. Most such nations have social or religious taboos that either preclude or limit domestic consumption. Table 6-9 lists those nations that have export potential but no competing domestic demand for the five export species.

CONCLUSIONS

Through our systems approach, we identify three critical elements in the economic feasibility of technology transfer: (1) relative cost of protein; (2) consumer acceptance and (3) the ability of the country in question to absorb the new technology. By assuming rankings for the values of these variables in each developing country, we are able to assess the probability of a successful transfer based upon a country's overall score. The results indicate that among our 90 developing countries Channel Catfish, Rainbow Trout, Mullet, Milkfish, Oysters and Tilapia can be transferred to a large number of these countries and have a high probability of successfully augmenting their food supplies. Mullet, Milkfish, Oysters, Rainbow Trout and Penaeus Shrimp have extensive potential for many developing coun-

Table 6-8. **World Price of Species Relative to Simulation Price**

Species	World Price[1]	Simulation Price Range[1]	Nations
Milkfish	$ .850 >	$.40 - .50	Burma*, Ivory Coast, Taiwan*
		.51 - .60	Mexico*, India, Thailand*, Brazil*, Ecuador
		.61 - .70	Morocco, S. Africa, Mainland China*, Indonesia*, Venezuela, Uruguay
		.71 - .85	Philippines*, Spain
Mullet	$ .260 >	.01	Egypt, Mozambique, S. Africa, Tunisia*, Cuba, Mexico, Bangladesh, Burma*, Mainland China*, Taiwan*, Hong Kong*, India*, Indonesia*, Iran, Korea R., Pakistan*, Philippines*, Syrian Arab R., Thailand*, Turkey, Argentina, Brazil, Chili, Columbia, Ecuador, Peru, Venezuela, Greece*, Portugal, Romania, Spain, Yugoslavia*, Costa Rica, El Salvador, Panama, Cyprus*, Israel*, Jordon, Lebanon, Albania, Uruguay
		.011 - .03	Sri Lanka, Poland
Oysters	$ .157 >	.11 - .07	Egypt, Ghana, Mozambique, S. Africa, Tunisia, Cuba, Mexico, Burma, Mainland China, Taiwan, India, Indonesia, Korea (R.), Pakistan, Philippines, Thailand, Brazil, Chili, Ecuador, Portugal, Costa Rica, El Salvador, Cyprus, Israel, Albania, Uruguay
		.07 - .156	Ghana, Syrian A.R., Turkey, Peru, Venezuela, Spain, Yugoslavia, Panama
Rainbow Trout	$ .380 >	$.08 - .12	Rhodesia, S. Africa, Mainland China, Taiwan*, Korea R., Turkey, Argentina, Columbia, Ecuador, Peru, Albania, Poland*
		.121 - .15	Venezuela, Greece, Hungary, Romania, Spain, Yugoslavia
		.21 - .23	Panama
Penaeus Shrimp	$1.170 >	$.29 - .39	Tunisia
		.40 - .60	Burma, Dahomey*
		.61 - .90	Thailand*
		.91 - 1.06	India*

1. Price expressed as U.S. dollars per kilogram of landed weight.
*Presently cultured to varying extents.
Source: World prices are taken from Table 6-7 and simulated prices from computer output. The simulated prices are those associated with the Aquaculture model in Chapter 5.

Table 6-9. Nations with Non-Competing Export Potential

Mullet	*Milkfish*	*Rainbow Trout*	*Penaeus Shrimp*	*Oysters*
Mozambique	Ivory Coast	Rhodesia	Tunisia	Egypt
S. Africa	Morocco	S. Africa		Ghana
Sri Lanka	Philippines	Turkey		Israel
		Argentina		Mozambique
		Venezuela		Pakistan
				S. Africa
				Syrian A.R.
				Tunisia
				Turkey

tries in terms of foreign exchange earnings. In some of these nations there is no competition with production for domestic consumption.

REFERENCES

Adelman, I. and Morris, C.T., *Society, Politics and Economic Development*, Baltimore: Johns Hopkins Press, 1962.

FAO Yearbook on Fisheries Statistics, Food and Agricultural Organization of the United Nations, Rome, 1971.

Harbison, F. and Myers, C.A., *Education, Manpower and Economic Growth*, New York: McGraw-Hill, 1964.

Unadjusted Costs of Protein per Kilogram by Sector by Country (c')

Note: Protein content of *sugar, fruits,* and *oils* is not significant. Less than 1 percent of edible commodity is protein. Only 1.4 percent of *vegetables* is protein. Therefore, these four groups should not be considered as possible intensive protein sources.

In other cases zeros mean there was not a significant amount consumed.

	Wheat	Rice	Potatoes & Starch	Pulses	Nuts & Seeds	Eggs	Milk
Algeria	.096	.065	.3916	.0699	.1713	.6623	.2321
Angola	.081	.101	.8216	.1195	.0629	.6607	.2749
Cameroon	.079	.094	.6298	.1200	.0701	.6607	.2609
Egypt	.077	.064	.5796	.0546	.0586	.6623	.2855
Ethiopia	.071	.720	.4589	.0427	.1197	.5803	.2507
Ghana	.082	.105	.5612	0.0000	.9356	.6574	.1762
Guinea	.081	.091	.8172	.1176	.0961	.6607	.2712
Ivory Co	.081	.104	.5732	.1446	.1153	.4412	.2843
Kenya	.060	.090	.6341	.1068	1.2957	.6618	.2067
Madagasc	.077	.416	.7535	.0796	.1481	66.4000	.3084
Malawi	.065	.074	.7268	.1314	.0878	.6627	.2741
Mali	.087	.321	.7131	.0697	.0737	.6607	.2274
Morocco	.079	.131	.3651	.0695	.5736	.6629	.2533
Mozambiq	.082	.103	.8121	.1223	.1064	.6574	.2847
Nigeria	.085	.071	.5562	.1080	.0892	.4968	.2687
Rhodesia	.081	.137	.3035	.1157	.0961	.6618	.2687
S. Africa	.075	.106	.3829	.1262	.0576	.6632	.2703
Sudan	.084	.068	.6792	.0704	.0766	.4412	.3146
Tanzani	.086	.113	1.2905	.0789	.8277	.6574	.2697
Tunisia	.197	.071	.3653	.0765	.2461	.5683	.2223
Uganda	.087	.093	.8249	.0254	.0515	.6574	.2809
Upper Vo	.072	.107	.7042	.3075	.0705	.6574	.2780
Zaire	.078	.111	1.0717	.1232	.1057	.6574	.3156
Zambia	.076	.131	.7212	.1210	.1713	.6607	.3320
Cuba	.107	.221	.6328	.1127	.1844	.9448	.3751
Dom Rep	.080	.210	.7321	.1272	1.3582	.9278	.4031
Guatemal	.070	.108	1.7297	.1111	.2010	.9254	.4230
Haiti	.082	.221	.8020	.1110	.6418	1.0113	.3663

Mexico	.110	.231	.8466	.1286	.2027	.9201	.3837
Afghanis	.082	.112	16.6000	.1126	.0769	.6607	.2796
Banglade	.096	.119	.4870	.0560	.2614	.6574	.3050
Burma	.101	.153	.4931	.1156	.1539	.5301	.3104
China Ma	.095	.112	.6691	.0453	.0377	.5896	.3289
China Ta	.096	.112	.6156	.0407	.0347	.6082	.2747
Hong Kon	.095	.112	.7455	.0791	.0660	.6180	.3363
India	.114	.118	.7952	.0412	.4853	.6574	.2967
Indonesa	.098	.112	.8279	.0312	.1131	.6607	.5446
Iran	.082	.112	.4963	.0827	.1975	.6618	.2818
Iraq	.082	.111	.3716	.0463	.1307	.4968	.2466
Khmer	.084	.112	.6607	.0000	.0000	.4968	.3289
Korea Dr	.095	.106	.5471	.0816	.0347	.5896	11.0000
Korea R	.072	.130	.3523	.0880	.0658	.5104	.0314
Malaysia	.095	.112	1.0326	.0358	1.4026	.5896	.3109
Nepal	.078	.100	.2395	.0840	.0000	.9910	.2940
Pakistan	.096	.119	.4870	.0560	.2614	.6574	.3050
Philippi	.089	.104	.9228	.0952	1.5581	.6031	.3125
Sau Arab	.084	.112	.4931	.0849	.0660	.4412	.2720
Sri Lank	.089	.158	.7090	.0613	.9334	.5524	.3125
Syr Ar R	.082	.114	.5175	.0560	.0923	.5524	.2533
Thailand	.066	.091	1.0395	.0394	.13922	.6031	.3295
Turkey	.086	.116	.8519	.1743	.8768	.5524	.3169
V Nam Dr	.094	.112	.8442	.0802	.0621	.6607	.3296
V Nam R	.097	.316	.6526	.0992	.3245	.6623	.3566
Yem Ar R	.082	.068	.4963	.0809	.0657	66.4000	.2177
Argentin	.033	.048	.4501	.0435	.2010	.3922	.3983
Bolivia	.084	.215	.6586	.0575	.0811	.8428	.3244
Brazil	.101	.200	.9774	.0972	.1887	.9278	.3785

	Wheat	Rice	Potatoes & Starch	Pulses	Nuts & Seeds	Eggs	Milk
Chili	.102	.211	.5641	.1330	.2010	.9059	.3811
Columbia	.112	.235	.9005	.1280	.2010	.9108	.3784
Ecuador	.113	.234	.9282	.1063	.2023	1.0105	.3810
Peru	.109	.212	.4877	.0982	.2010	.8088	.4417
Venezuel	.088	.230	.6752	.1170	.4020	.8999	.3740
Greece	.083	.181	.2980	.1052	.1976	.4774	.2441
Hungary	.074	.136	.2783	.1279	.2227	.4353	.2641
Poland	.089	.197	.2940	.1183	0.0000	.3064	.2946
Portugal	.117	.167	.2348	.1348	.3614	.5265	.3141
Romania	.089	.167	.2938	.1196	.1205	.4915	.2960
Spain	.102	.209	.2826	.1135	.4942	.4789	.2859
Yugoslav	.087	.167	.2966	.1091	.2299	.4915	.3046
Chad	.069	.101	.7436	.1191	.0706	.6607	.2325
Congo	.080	.101	1.2242	.1126	.0746	.6607	.2845
Dahomey	.072	.113	.5145	.1163	.0736	.6574	.3317
Gabon	.078	.107	1.0005	.1308	.0897	.6574	.3161
Lib Ar R	.089	.114	.4955	.0427	.0710	.5524	.2597
Niger	.057	.113	.6648	.1193	.0679	.6618	.2557
Si Leone	.078	.107	1.1636	.0674	.1691	.6607	.2253
Cos Rica	.064	.308	.8853	.1272	0.0000	1.1004	.2172
El Salva	.052	.148	1.2266	.1091	.2010	1.0720	.4010
Honduras	.051	.115	1.1789	.1090	.3861	1.0124	.3813
Nicaragu	.053	.133	1.0923	.1091	.2010	1.0123	.3990
Panama	.080	.234	.7239	.0907	.9442	.9278	.3858
Cyprus	.098	.104	.4977	.1737	.4700	.5899	.3058
Israel	.080	1.227	.4931	.1126	.0831	.6117	.3166
Jordon	.084	.102	.4972	.0618	.3134	.5896	.2427
Laos	.100	.112	.8776	.0834	.0657	.6631	.3297
Lebano	.085	.114	.5285	.0831	.1033	.6633	.2879
Albania	.089	.191	.3161	.1195	0.0000	.4208	.2951
Iceland	.089	.018	.2936	1.0416	0.0000	.5264	.3034
Paraguay	.099	.239	1.0059	.0887	.0812	.9400	.3770
Uruguay	.085	.216	.6191	.1487	0.0000	.8899	.3770

	Coarse Grains	Beef	Mutton	Pork	Poultry	Fin Fish	Shell Fish
Algeria	.077	.395	.502	0.000	.379	.134	2.280
Angola	.036	.344	.612	.487	.292	.102	0.000
Cameroon	.083	.387	.474	.485	.285	.081	2.280
Egypt	.072	.693	.709	0.000	.253	.150	0.000
Ethiopia	.073	.390	.521	0.000	.384	0.000	0.000
Ghana	.035	.395	.434	.452	.493	.095	2.280
Guinea	.035	.415	.641	.452	.365	.095	0.000
Ivory Co	.038	.378	.518	.473	.344	.098	2.280
Kenya	.041	.302	.478	.452	.365	.123	0.000
Madagasc	.036	.390	.320	.443	.448	.097	.056
Malawi	.058	.422	.434	.481	.438	.063	0.000
Mali	.065	.368	.503	.452	.425	.083	0.000
Morocco	.079	.372	.502	4.970	.501	.197	0.000
Mozambiq	.036	.369	.612	.481	.471	.082	2.280
Nigeria	.065	.331	.546	.237	.305	.081	2.180
Rhodesia	.036	.320	.492	.487	.340	.103	0.000
S Africa	.036	.431	.587	.443	.425	.143	.147
Sudan	.343	.651	.693	0.000	.303	.167	0.000
Tanzani	.051	.379	.477	0.000	.406	.051	.147
Tunisia	.076	.355	.521	0.000	.501	.174	2.280
Uganda	.029	.385	.424	0.000	.342	.096	0.000
Upper Vo	.064	.364	.474	.473	.425	.075	0.000
Zaire	.036	.422	.612	.473	.299	.097	0.000
Zambia	.036	.391	.434	.904	.505	.106	0.000
Cuba	0.000	.602	.844	.649	.377	.098	.326
Dom Rep	.081	.551	0.000	.614	.050	.078	0.000
Guatemal	.126	.544	0.000	.614	.406	0.000	.109
Haiti	.085	.642	.679	.691	.471	.132	0.000

	Coarse Grains	Beef	Mutton	Pork	Poultry	Fin Fish	Shell Fish
Mexico	.093	.648	.609	.771	.425	.128	.207
Afghanis	.157	.674	.694	.000	.493	.000	.000
Banglade	.084	.741	.679	.000	.471	.141	.000
Burma	.081	.597	9.280	.756	.505	.091	.147
China Ma	.087	.674	.599	.772	.455	.199	.000
China Ta	.120	.603	.000	.681	.380	.089	.141
Hong Kon	.093	.552	.844	.473	.385	.275	.000
India	.096	.850	.599	.000	.471	.075	.207
Indonesa	.087	.616	.844	.738	.247	.071	.000
Iran	.092	.642	.771	.000	.406	.000	.207
Iraq	.108	.648	.735	.000	.334	.075	.000
Khmer	.625	.648	.000	.768	.501	.142	.000
Korea Dr	.083	.527	.844	.710	.305	.253	.000
Korea R	.186	.210	.000	.214	.129	.120	.200
Malaysia	.096	.790	.884	.694	.385	.198	.479
Nepal	.087	1.140	.884	7.750	.379	.147	.000
Pakistan	.084	.741	.679	.000	.471	.141	.000
Philippi	.089	.465	.000	.458	.300	.117	.278
Sau Arab	.117	.571	.733	.000	.493	.175	.415
Sri Lank	.092	.684	.844	.705	.471	.070	.207
Syr Ar R	.096	.590	.739	.000	.406	.221	.000
Thailand	.075	.688	.000	.770	.425	.156	.197
Turkey	.083	.645	.693	.000	.740	.174	.000
V Nam Dr	.096	.648	.000	.748	.423	.195	.000
V Nam R	.100	.613	.000	.735	.379	.198	.434
Yem Ar R	.114	.719	.825	.000	.126	.074	.000
Argentin	.041	.630	.740	.744	.425	.113	.207
Bolivia	.087	.622	1.380	.354	.740	.111	.000
Brazil	.081	.587	.229	.076	.050	.068	.109
Chili	.183	.656	.597	1.788	.343	.090	.278
Columbia	.088	.617	.092	7.045	.510	.090	.207

Ecuador	.086	.660	.714	.728	.340	.113	.109
Peru	.123	.567	.632	.769	.410	.109	.415
Venezuel	.096	.626	.844	.770	.422	.081	.415
Greece	.107	.814	1.109	.731	.438	.114	0.000
Hungary	.085	.772	1.057	.624	.372	.151	0.000
Poland	.115	.880	1.376	.765	.398	.198	0.000
Portugal	.111	1.067	1.078	.670	.386	.099	.074
Romania	.110	.792	1.244	.826	.435	.193	0.000
Spain	.156	.866	1.233	.870	.420	.134	0.000
Yugoslav	.043	.867	1.001	.804	.429	.090	0.000
Chad	.069	.374	.477	0.000	.471	.128	0.000
Congo	.034	.364	.612	.710	.305	.091	0.000
Dahomey	.039	.377	.434	.490	.294	.102	.023
Gabon	.037	.392	.641	.904	.288	.104	0.000
Lib Ar R	.060	.592	.676	0.000	.406	.101	0.000
Niger	.041	.363	.472	.452	.379	.217	0.000
Si Leone	.052	.395	.612	.452	.320	.099	0.000
Cos Rica	.174	.538	0.000	.698	.425	.100	.089
El Salva	.110	.578	0.000	.489	.340	.084	.207
Honduras	.106	.293	0.000	.690	.292	0.000	.147
Nicaragu	.118	.553	0.000	.667	.379	0.000	.075
Panama	.092	1.049	0.000	.957	.747	.123	.102
Cyprus	.094	.607	.700	.692	.409	.099	0.000
Israel	.075	.623	.728	0.000	.413	.151	0.000
Jordon	.075	.616	.665	0.000	.365	.089	0.000
Laos	.092	.674	0.000	.768	.410	.217	2.280
Lebanon	.081	.589	.661	0.000	.448	.135	0.000
Albania	.111	.810	1.144	.721	.340	.327	0.000
Iceland	.090	.870	1.204	0.000	.308	.110	0.000
Paraguay	.101	.621	.442	.782	.407	.207	0.000
Uruguay	0.000	.655	.722	.771	.340	.090	.207

Adjusted Costs of Protein by Sector by Country (c''): Code A*

KEY:

R	=	Rice
CG	=	Coarse Grains
FF	=	Fin-Fish (wild stock)
SF	=	Shellfish (wild stock)
W	=	Wheat
PL	=	Poultry
B	=	Beef
PK	=	Pork
M	=	Mutton
CAR	=	Indian Carp
CC	=	Channel Catfish
TL	=	Tilapia
E	=	True Eel
TR	=	Rainbow Trout
ML	=	Mullet
MK	=	Milkfish
Y	=	Yellowtail
PS	=	Penaeus Shrimp
O	=	Oysters
MU	=	Mussels
MB	=	Macrobrachium
BLA	=	Blue-Green Algae
WC	=	Walking Catfish

*Per 10 kilograms
Source: Computer Aquaculture Model

Country						
Algeria	ML .05	R .91	CG 1.83			
Angola	ML .07	R 1.41	FF 1.45	W 2.03		
Cameroon	ML .06	O 1.29	R 1.31	CG 1.97	W 1.98	TL 2.83
Egypt	ML .08	R .88	CG 1.71	W 1.92		
Ethiopia	ML .05	O 1.14	TR 1.23	CG 1.73	W 1.78	TL 3.40
Ghana	ML .07	O .07	CG .82	FF 1.36	R 1.46	
Guinea	ML .05	CG .84	R 1.27	FF 1.37		
Ivory Coast	ML .05	CG .90	O .91	FF 1.39	R 1.45	
Kenya	ML .06	TR .94	CG .96	R 1.25		W 1.49
Madagascar	ML .05	O .54	SF .80	CG .86	FF 1.39	
Malawi	FF .90	R 1.02	TL 1.73			
Mali	FF 1.19	CG 1.55				
Morocco	ML .06	TR 1.66	R 1.82	CG 1.87		
Mozambique	ML .06	CG .85	FF 1.17	R 1.43		O 2.02
Nigeria	ML .06	R .99	FF 1.16	O 1.23	CG 1.54	
Rhodesia	CG .86	TR .94	FF 1.47			
South Africa	ML .06	TR .83	CG .87	R 1.48		
Sudan	ML .05	R .95	W 2.11			
Tanzania	ML .05	FF .07	O .41	CG 1.21	R 1.52	TL 1.98
Tunisia	ML .05	O .45	R .98	CG 1.80	TL 2.56	FF 2.49
Uganda	CG .69	R 1.29	FF 1.37			
Upper Volta	FF 1.07	R 1.48	CG 1.53			
Zaire	CG .86	FF 1.39	R 1.54			

Country						
Zambia	CG .85	FF 1.52	R 1.82			
Cuba	ML .07	FF 1.41	W 2.67	R 3.07	O 3.75	
Dominican Republic	ML .07	PL .52	FF 1.12	CG 1.92		
Guatemala	ML .07	R 1.50	SF 1.55	TR 1.72	W 1.74	
Haiti	ML .07	FF 1.89	CG 2.03	W 2.05		
Mexico	ML .07	TR .98	FF 1.84	CG 2.22		
Afghanistan	TR .74	R 1.56				
Bangladesh	ML .06	R 1.65	CG 2.01	FF 2.01		
Burma	ML .05	O .67	FF 1.30	TR 1.49	CG 1.93	SF 2.10
China Mainland	ML .06	TR .80	R 1.55	O 1.62	CG 2.07	W 2.38
China Taiwan	ML .06	TR .87	FF 1.27	R 1.56		
Hong Kong	ML .33	R 1.56	TR 1.98	CG 2.20		
India	ML .06	FF 1.07	O 1.19	R 1.64	CG 2.29	
Indonesia	ML .06	FF 1.02	R 1.55	O 1.93	CG 2.08	
Iran	ML .06	R 1.56	TR 1.73	W 2.05		
Iraq	ML .12	FF 1.07				
Khmer	ML .06	R 1.55	O 1.60	FF 2.04	W 2.11	
Korea (DP)	ML .07	TR .83	R 1.47	CG 1.98		
Korea (R)	ML .07	TR .85	FF 1.72	R 1.81		
Malaysia	ML .06	R 1.55	O 1.76	CG 2.30	W 2.38	FF 2.83
Nepal	TR .75	R 1.39				
Pakistan	ML .07	R 1.65	CG 2.01	FF 2.01	W 2.39	O 3.13
Philippines	ML .06	R 1.44	FF 1.68	CG 2.16		

Saudi Arabia	ML	R	W		
	.17	1.56	2.11		
Sri Lanka	ML	FF	R	CG	
	.14	1.00	2.20	2.20	
Syrian Arab Republic	ML	R	W		
	.06	1.59	2.05		
Thailand	ML	O	R	W	CG
	.05	1.08	1.27	1.66	1.77
Turkey	ML	TR	R		
	.08	1.02	1.62		
Viet Nam DR	ML	R	O	CG	W
	.06	1.56	1.57	2.30	2.34
Viet Nam R	ML	O	CG	W	FF
	.06	2.02	2.38	2.42	2.82
Yemen Arab Republic	O	ML	R	FF	TL
	.03	.05	.95	1.05	1.13
Argentina	ML	R	TR	W	
	.08	.67	.68	.82	
Bolivia	TR	FF			
	.75	1.59			
Brazil	ML	FF	SF	CG	
	.07	.97	1.55	1.92	
Chile	ML	TR	FF	W	
	.07	.97	1.28	2.54	
Columbia	ML	TR	FF	CG	
	.08	.85	1.29	2.09	
Ecuador	ML	TR	SF	FF	
	.06	.84	1.55	1.61	
Peru	ML	TR	FF	W	CG
	.07	.97	1.55	2.72	2.93
Venezuela	ML	FF	TR	CG	W
	.08	1.15	1.32	2.28	2.71
Greece	ML	TR	FF	W	
	.10	1.14	1.63	2.07	
Hungary	TR	W			
	1.13	1.86			
Poland	ML	TR			
	.17	1.07			
Portugal	ML	SF	FF	TR	
	.07	1.06	1.42	1.55	
Romania	ML	SF			
	.09	1.13			
Spain	ML	TR	FF		
	.08	1.16	1.92		
Yugoslavia	ML	CG	TR	FF	
	.09	1.03	1.09	1.28	

Chad	R	CG	W	TL	
	1.41	1.64	1.73	1.80	
Congo	ML	O	CG	FF	R	TL
	.05	.76	.80	1.30	1.41	1.45
Dahomey	ML	SF	O	CG	FF	R
	.04	.32	.64	.92	1.45	1.57
Gabon	ML	CG	R	FF	
	.23	.89	1.48	1.48	
Libyan Arab Republic	ML	CG	FF	
	.15	1.43	1.44	
Niger	CG	W	R	TL	
	.99	1.42	1.57	1.93	
Sierra Leone	ML	O	CG	FF	R
	.05	.41	1.24	1.41	1.49
Costa Rica	ML	TR	SF	FF	W	O
	.07	1.68	1.28	1.43	1.61	3.30
El Salvador	ML	FF	W	TR	
	.07	1.21	1.31	1.69	
Honduras	ML	W	R	TR	
	.07	1.28	1.59	1.70	
Nicaragua	ML	W	TR	R	O
	.07	1.35	1.67	1.85	2.77
Panama	ML	SF	FF	TR	
	.08	1.46	1.76	1.80	
Cyprus	ML	FF	R	CG	
	.08	1.42	1.45	2.24	
Israel	ML	CG	W	
	.12	1.77	1.99	
Jordan	ML	FF	R	CG	
	.06	1.28	1.41	1.77	
Laos	R	CG	
	1.56	2.18	
Lebanon	ML	R	CG	FF	
	.13	1.59	1.92	1.94	
Albania	ML	TR	W	CG	
	.07	.91	2.21	2.63	
Iceland	FF	CG	
	1.57	2.14	
Paraguay	CG	W	FF	
	2.41	2.47	2.96	
Uruguay	ML	FF	W	SF	
	.07	1.29	2.14	2.96	
Australia	ML	CG	FF	TR	
	.12	1.46	1.46	1.47	
Canada	ML	CG	W	FF	
	.17	.94	1.26	1.55	

Czechoslavakia	ML	TR	W	R	
	.12	1.34	2.22	2.72	
Denmark	ML	TR	R	FF	
	.13	1.49	1.64	1.74	
France	ML	TR	FF	W	
	.10	1.09	1.38	1.68	
Germany	ML	TR	FF	W	
	.12	1.38	1.38	2.42	
Italy	ML	TR	FF	R	
	.13	1.45	1.68	1.76	
Japan	ML	TR	R	FF	
	.13	1.19	1.62	1.79	
New Zealand	ML	TR	CG	W	FF
	.11	1.29	1.62	1.64	1.92
Norway	ML	FF	TR	W	
	.17	1.80	1.86	2.05	
United Kingdom	ML	TR	W	FF	
	.14	1.62	1.62	1.96	
United States	ML	W	FF	TR	CG
	.20	1.62	1.70	2.30	2.33

✳ *Appendix 6-C*

Technology Acceptance Indicators

Country	Rank Invest. Share of GNP	Harbison-Myers Index	Extent of Literacy	Level of Modernization Economic	Commitment to Econ. Growth	Degree, Administrative Efficiency	Extent of Mass Communication	Degree of National Integration	Technological Acceptance Score
Algeria		C	C-	B	B-	B+	B-	C	III
Angola							B		III
Cameroon	43	C-	D-	C	B-	B-	B	C	III
Egypt	45	A-							II
Ethiopia	46	D-	D-	D	C	C+	D-	C	IV
Ghana	48	B-	C-	C		B+	B-	B-	II
Guinea		D	D	D	B-	C+	D-	D	IV
Ivory Coast	23	D	D	D	B-	B-	D+	D	IV
Kenya	16	D+	C-		B-	A-	C-	D+	III
Madagascar	40								IV
Malawi	26	D-	D-	D	C	A-	D-	D+	III
Mali									III
Morocco	33	C	D+		C	B-	B-	C	III
Mozambique				C+	C				II
Nigeria	36	D+	D	C-	B+	B+	D+	D+	III
Rhodesia	30	D	C	C+	C	B-	B-	C	II
South Africa	5	A-	B-	A-	A-	A-	B+	D	I
Sudan	47	C-	D	C	B-	B+	D	C-	III
Tanzania	13	D	D	C					III
Tunisia	9	C+	C-	C	A-	A-	B-	B	
Uganda		D+	C-	C	B-	A-	C-	D	III
Upper Volta	50	D+							V
Zaire	2								III
Zambia	14	D	C	C	C	A-	D+	D	III
Cuba		B+			C				II
Dominican Republic	38	C	B-	B-	C	C+	B-	B	III
Guatemala	41	C	C	C	C	C+	B-	C	III

Country	No.	1	2	3	4	5	6	7	8	
Haiti		D+		B	B	A−	A−	A−	A	V
Mexico	22	B		D	D	B−	C+	D	D	I
Afghanistan		D−								IV
Bangladesh		B								IV
Burma	49	C	B	C−	B−	B−	C−	C−	C	II
China Mainland	7	C+	B				A			II
China Taiwan	19	A	A							I
Hong Kong	35	A	C−				C−			I
India		B+	B+	A+	A+	A+		C−	D+	I
Indonesia	11	C	C	C+	B−	C	C−	C−	B−	II
Iran	37	C+		C+	B−	B−	C	B−	C	II
Iraq	29	B	A−	C−	C	C−	C−	C−	C	III
Khmer			B−			C−				III
Korea D.P.	4	A	A	B−	C	C	A	A−	B	III
Korea, R.		B	B			C−				II
Malaysia		C−						A−		III
Nepal	42	B	D−	C−	C	D	D	D−	C−	III
Pakistan	21	A+	D+	B−	A−	C	D+	C	C+	II
Philippines	28	D−	A	B+	B−	B−	A	A−		I
Saudi Arabia	25	B				B−		C+		III
Sri Lanka	32	A	B+	B−	C	B−	B−	B+	B	II
Syrian Arab Republic	8	B+	C	B−	C	B−	C	C−	B	II
Thailand	18	B	A−	B−	B−	B−	A−	B−	C	I
Turkey			B−	A−	B−			B−	A	I
Viet Nam D.R.						C				III
Viet Nam R.	17	C	C−	B−	C	D	C−	C+	C	III
Yemen Arab R.	39	D−		C+	C	A−	D−	D−	C−	V
Argentina	31	A+	A+	B−	B−	A−	B−	A−	A	I
Bolivia	34	C	C+	C+	B−	D+	C+	B−	D	III
Brazil	12	B−	A	B+	B+	A−	A	B	B	II
Chili		A	A	A−	B+	A−	A	A−	A	I
Columbia		B−	B+	A−	B+	B	B+	B+	A	I
Ecuador	27	B−	B+	C+	C	B−	B+	B	B−	II

Country	Rank Invest. Share of GNP	Harbison-Myers Index	Extent of Literacy	Level of Modernization Economic	Commitment to Econ. Growth	Degree, Administrative Efficiency	Extent of Mass Communication	Degree of National Integration	Technological Acceptance Score
Peru	44	B	B	B	B−	B−	A−	C+	II
Venezuela	6	A	B+	A−	A	A−	A−	A	I
Greece	3	A	A	A−	A−	A+	A	A	I
Hungary	15	A							I
Poland	20	A							I
Portugal	24	A−							I
Romania									I
Spain	10	B+							I
Yugoslavia	1	A							I
Chad	66	D−	D−	D	B−	C+	D−	D	IV
Congo		D							IV
Dahomey	60	D	D+	D	B−	B−	D−	D	IV
Gabon	51	C−	D	D	B+	B+	B−	C	III
Libyan Arab Republic	59	C	D+	D	C	C−	C+	C	III
Niger	68	D−	D−	D	B+	C−	D−	D	IV
Si Leone	61	D+	D	C−	B−	B−	D+	D	III
Costa Rica	56	A	A	B	C	A−	A−	A−	I
El Salvadore	65	C+		B−	C	B+	B+	B	II
Honduras	58	C		B−	C	C+	B−	B−	III
Nicaragua	62	C	B−	B	C	B−	B+	B−	III
Panama	55	A	A−	C+	C	B−	A	B	II
Cyprus	54	A	B+	A−	C	B−	A+	C	I
Israel	53	A+	A+	A+	A+	A+	A+	B+	I
Jordon	64	B+	C−	B−	B−	C−	C+	B	II
Laos		D		D	C	C−	D−	C	III
Lebanon	57	B−	B	A−	C	B−	A−	B	II
Albania	52								II
Iceland									I
Paraguay	63	B−	A−	C−	C	C−	B−	B−	I
Uruguay	67	A	A+	A−	C	A−	A+	A	III

※ *Chapter Seven*

Land Suitability, Production, Employment, and Foreign Exchange Earnings

Even where environmental and economic conditions are favorable for aquaculture production, sufficient land may not be available inland or sufficient coastline seaward. We examine this issue in a very limited way in this chapter. Given our environmental variable we also estimate production per hectare for the selected species. The actual production levels would, of course, depend upon the total area devoted to such production. From the man-year input data for our model operations we also are able to compare the labor-intensiveness of aquaculture production with that of agricultural production among the potential adopters. The production estimates enable us to estimate foreign exchange earnings per hectare where trade potential exists. Finally, all these (admittedly crude and tentative estimates) are assessed as they might relate to the small farmer and hence to the income distribution.

LAND SUITABILITY FOR INLAND AND COASTAL AQUACULTURE PRODUCTION

All of the species selected for transfer for either low-cost domestic protein (e.g., Mullet, Milkfish, Channel Catfish, Rainbow Trout, Tilapia) or for potential export earnings (e.g., Mullet, Rainbow Trout, Milkfish, Penaeus Shrimp) can be raised in estuaries. This suggests that, for the most part, very little land would need to be diverted from agricultural production for aquaculture production. Thus the question regarding land use depends first upon general availability of suitable coastline and specifically upon alternative coastline usage such as for recreation, tourism and commercial fishing of wildstocks. The second or opportunity-cost aspect of this question goes well beyond the scope of this study. Some species may have the potential to serve a dual purpose, both as a source of domestic protein and foreign exchange.

Three of the species selected from the highest criteria set, *Channel Catfish*, *Tilapia* and *Rainbow Trout*, can also be farmed inland. Our data on Rainbow Trout, however, are based upon a *saltwater* operation. Unfortunately, therefore, our data cannot reveal much on inland productivity or labor-intensiveness. Our Channel Catfish and Tilapia data are based upon inland operations and for comparable operations are applicable to production and employment considerations.

A rough estimate of the percent of total land area in each nation suitable for inland fishery production for these three species was made by our biologists and physical geographers from FSU's Florida Resources and Environmental Analysis Center. These data reduce slightly the non-estuarine production potential for Tilapia, but not Channel Catfish. In Algeria only 10 percent of the land is suitable for Tilapia and in the Sudan, only 15 percent. To the extent Trout is cultured inland, the implication is that fish production will be substituted for some agricultural production and employment.

Not all the land shares suitable for production are available. Some are currently in use for agricultural production and for cities and towns. Table 7-1 shows the comparison of suitable with available inland land for Channel Catfish and Tilapia in selected nations. Agricultural employment and shares of urban population were used to adjust usable land shares. Presumably in these countries land could be used for Channel Catfish and Tilapia culture without detracting from production or employment in other sectors.

Table 7-1. Shares of Total Land Available for Channel Catfish and Tilapia, Selected Nations

	Percent Total Land Areas	
Country	*Unadjusted*	*Adjusted*
Cameroon	100%	55%
Burma	90	35
Indonesia	75	53
Khmer	95	69
Malaysia	100	69
Thailand	95	66

It should be remembered that our criteria for species-country selection had as one of its components the potential for substitution of aquaculture for current protein producing sectors. Even where land already is in agricultural use it would still be to the advantage of a country from a pure protein standpoint to substitute aquaculture for agriculture use based upon the results reported in Chapter 6.

Let us now turn our attention to the saltwater species and hence the availability of coastline for their culturing. You will recall that species with the highest potential for transfer and modeled from saltwater experience were Mullet, Milkfish, Mussels and Oysters. (Rainbow Trout normally is not cultured as a saltwater species.) To these species could be added Yellowtail, Penaeus Shrimp, and Macrobrachium. The hectares of coastline suitable for fish production for our sample nations are provided in Table 7-2. This listing should not be confused with the *specific* countries found to have potential for the transfer of technology of a particular type. (A cross-listing can be made by comparing Table 7-2 with Tables 6-4 and 6-5.)

The availability of coastline constrains potential production only slightly. Malawi, which has an average potential for the transfer of Channel Catfish, exhibits that potential in freshwater areas only. Likewise, in Bolivia (average), Paraguay (average), Malawi, Mali, Zaire, and Zambia—where a potential for freshwater Tilapia was found—no coastline is available for the transfer of such operations into saltwater. (Remember that Tilapia has a high tolerance for saltwater.) In Hungary Rainbow Trout could only be cultured in freshwater areas. Otherwise, our findings of Chapter 6 remain intact. Still, a caveat is in order. These data represent the absolute upper limit to production potential. Again, we do not know what actual share of such coastline has competing uses and exploration of this question is beyond the scope of this inquiry.

TENTATIVE PRODUCTION AND EMPLOYMENT IMPLICATIONS

The data on production and employment apply only to the type and size of operation of our sample firm for each species. Any further extrapolation may not be valid because of the unavailability of *detailed* information on the topography of each country.

Table 7-3 provides data on man-years per hectare for those species of highest potential for transfer for domestic production and for Penaeus Shrimp which has only export potential. Three of the same species (Mullet, Milkfish and Rainbow Trout) also have export potential (see Table 6-8).

Table 7-2. Hectares of Coastline Suitable for Coastal Fish Production by Country

Country	Hectares	Country	Hectares	Country	Hectares	Country	Hectares
Algeria[a]	777,600	Chad	0	Saudi Arabia[d]	1,814,400	Mexico	9,201,600
Angola	777,600	Congo	777,600	Thailand	2,721,600	Greece	1,944,000
Cameroon	518,400	Dahomey	194,400	Turkey[e]	6,000	Hungary	0
Egypt	842,400	Gabon	2,073,600	North Vietnam	1,036,800	Poland[h]	0
Ethiopia	453,600	Libya	907,200	South Vietnam	3,628,800	Portugal	1,360,800
Ghana	972,000	Niger	0	Yemen	103,680	Rumania[i]	0
Guinea	583,200	Sierra Leone	1,036,800	Cypress	103,680	Spain[k]	3,110,400
Ivory Coast	972,000	Afghanistan	0	Israel	518,400	Yugoslavia	1,360,800
Kenya	1,296,000	Bangladesh	1,944,000	Jordan	12,960	Costa Rica	1,069,200
Madagascar	3,868,000	Burma	5,184,000	Laos	0	El Salvador	583,200
Malawi	0	China	4,924,800	Lebanon	466,560	Honduras	1,749,600
Mali	0	Taiwan	1,944,000	Bolivia	0	Nicaragua	1,555,200
Morocco	1,944,000	Hong Kong	32,400	Brazil	8,294,400	Panama	3,175,200
Mozambique	4,276,800	India	8,164,800	Argentina[f]	3,888,000	Paraguay	0
Nigeria	2,073,600	Indonesia	23,587,200	Chile[g]	5,433,200	Uruguay	1,166,400
Rhodesia	0	Iran	2,851,200	Columbia	2,912,000	Albania	583,200
South Africa	3,240,000	Iraq[b]	0	Ecuador	1,360,800	Iceland[l]	1,594,080
Sudan	907,200	Khmer	388,800	Peru	2,851,200	Sri Lanka	7,500
Tanzania	1,555,200	Korea Republic	583,200	Venezuela	5,702,400	Syrian Arab Republic[m]	1,000
Tunisia	1,036,800	North Korea[c]	1,944,000	Cuba	4,276,800		
Uganda	0	Malaysia	1,944,000	Dominican Republic	1,166,400		
Upper Volta	0						

Zaire	0	Nepal	0	Haiti	1,296,000
Zambia	0	Pakistan	1,166,400	Guatemala	777,600
		Philippines	5,832,000		

a Zero available for Yellowtail.
b 64,800 hectares available for Mullet and Yellowtail.
c Zero available for Milkfish.
d Zero available for Milkfish.
e Zero available for Milkfish, and 1,555,200 hectares available for Mullet.
f Zero available for Milkfish and Yellowtail.
g Zero available for Milkfish.
h 777,600 hectares available for Mullet, Oysters, and Mussels.
i 583,000 hectares available for Mullet and Seaweed.
k Zero available for Macrobrachium.
l Zero available for Milkfish, Yellowtail, Macrobrachium, and Penaeus Shrimp.
m Zero available for Milkfish.

Source: Dr. Randy Martin of the Biological Sciences Department and Mr. Robert Glassen of the FSU Center for Resources and Environmental Analysis.

Table 7-3. Labor Usage: Aquaculture Models of Highest Potential

Species	Man-Years per Hectare
Channel Catfish	.0370
Rainbow Trout	29.444
Mullet	1.114
Milkfish	.066
Mussels	22.75
Oysters	1.000
Tilapia	.186
Penaeus Shrimp	.654

Source: Table 3-1.

In order to compare relative labor-intensiveness we also estimated the man-years per hectare usage in agriculture in the sample nations. A sample of these results is provided in Table 7-4.

Finally, from our simulation model of Chapter 5 we estimated the gross production weight per hectare for the selected species. The following formula applies:

$$\text{Production per Hectare} = \frac{EQ}{D},$$

where Q = output in kilograms, E = non-market environmental variable, and D = size of model operation in hectares. The value of E varies by nation.

The simulated production per hectare of *Mullet* exceeds 103,300

Table 7-4. Labor Usage in Agriculture: Selected Countries

Country	Man-Years per Hectare
Guatemala	.8
Mexico	.3
China P.R.	2.1
India	.9
Columbia	.8
Venezuela	.2
Honduras	.7
Peru	.7
Brazil	.4

kilograms per hectare for most nations where transfer is favorable. In the sample operation for Mullet, at a labor to land ratio of 1.114 man-years per hectare, this operation is more labor-intensive than agriculture in general in several nations. In any case, since the coastal operation may be complementary in employment, the 1.114 man-years can be viewed generally as the potential net employment gain per hectare for the operation of this type of enterprise.

Our *Milkfish* operation was less labor-using. Some .066 workers per hectare were employed which compares favorably to the agricultural labor-hectare ratio only in such nations as *Taiwan*, *Malaysia* and *Uruguay*. Again, however, we can view this species as net employment-creating in *coastal* operations. Production per hectare for Milkfish is simulated at 1500 kilograms for all selected nations (i.e., the environment coefficient, E, equals unity).

Our *Channel Catfish* operation would generate 2016 kilograms of output per hectare in all nations except Mainland China where production would be 1613 kilograms. This particular operation uses little labor, but nonetheless would be a net employment producer on the coastline and would compare favorably with agriculture in Malaysia.

The sample data on *Rainbow Trout* generates an extremely large yield of between 1,252,380 and 626,253 kilograms per hectare. Also, this is a very labor intensive operation whose labor-land ratio exceeds the agriculture-labor to land ratio in all the producing nations.

The enterprise producing *Mussels* would generate a simulated 44,211 kilograms of output in Khmer. The potential for net employment creation is great, given the labor-intensive nature of the simulated operation.

The *Penaeus Shrimp* operation yields between 1052 and 1803 kilograms per hectare and would be relatively labor-intensive in Tunisia, Burma, Dahomey and India. This type of operation could add substantially to net employment in coastal regions.

Again, we urge caution. These results are to be viewed as only representative for the specific type of operation from which our data was derived.

Our *Oysters* operation yields 90,729 kilograms of output in most nations. But output runs as low as 27,225 kilograms in Iran and as high as 108,873 kilograms in Chile. This is a labor-intensive operation relative to average labor-land ratios in all nations except for Egypt, China (P.R.), Hong Kong, Indonesia, Khmer, Korea (D.R.), Korea (R.), Saudi Arabia, Thailand, Vietnam (D.R.), Vietnam (R.), El Salvador, Lebanon and Albania.

Finally, *Tilapia* operations yield only 344 kilograms per hectare in virtually all nations. It should be noted, however, that our data were derived from cage culture production. This qualification also must be kept in mind when we note that such culture is extremely labor-intensive relative to the amount of coastline that a cage occupies.

FOREIGN EXCHANGE EARNINGS

Foreign exchange earnings were estimated for Mullet, Penaeus Shrimp, Milkfish, and Rainbow Trout. The values are based upon the competitive world prices (in U.S. dollars) from Table 6-7. The simple formula is:

$$\begin{pmatrix} \text{Foreign Exchange} \\ \text{Earnings per} \\ \text{Hectare} \end{pmatrix} = \begin{pmatrix} \text{World} \\ \text{Price} \end{pmatrix} \times \begin{pmatrix} \text{Production} \\ \text{per} \\ \text{Hectare} \end{pmatrix}$$

Again, values are expressed in terms of revenue per hectare and are FOB values. The simulations apply to all those countries earlier identified as having foreign trade potential (see Table 6-8).

The foreign exchange value per hectare of *Rainbow Trout* is very high because of the high-yield value of the operation sampled. The simulated earnings range from $237,976 per hectare in Panama to $571,076 per hectare in a nation like Columbia.

The next highest foreign exchange earner in the simulations is *Mullet* where earnings are $26,858 per hectare in all nations except the Libyan Arab Republic.

Penaeus Shrimp generates from $1052 per hectare in a nation like Burma to $1803 in a nation such as Dahomey.

Our *Milkfish* generates $1275 per hectare in foreign exchange in all nations where it would be suitable to produce for export.

The value of foreign exchange for Oysters is estimated as $14,244 per hectare in most nations having export potential.

THE SMALL FARMER AND THE INCOME DISTRIBUTION

In agriculture a necessary condition for production is to own or otherwise have access to land. In many developing nations the distribution of such rights is extremely unequal. But, as our small-scale models indicate, aquaculture enterprises need access to only a fraction of a hectare with some species and one hectare operations are not uncommon. Capital markets in developing nations also are imperfect and the cost of capital to the peasant is often prohibitive. Hence, a small-scale enterprise again is favorable for making the income distribution less unequal. Smaller operators also are more likely to use labor-intensive methods of production and thus further diminish income inequalities.

The aquaculture operations in our sample are of relatively small scale. Thus, the results appear to be promising for the small farmer. An expansion in the number of such smallish producing units should tend to make the income distribution more equal. Moreover, since most of the species selected for transfer are for coastline production, aquaculture production should add to employment of labor and hence tend to make the income distribution more equal. Nonetheless, even where small-scale production is possible, governmental subsidies may be required initially in order to stimulate the new technique.

CONCLUSIONS

In general we can conclude that suitable land availability is not a significant constraint on the expansion of aquaculture activities. This is the case both inland and seaward. If unit protein costs are favorable relative to those of other food products, some substitution of agricultural land use for aquacultural use would be warranted economically. The case cannot be stated quite as strongly for coastline production because of our lack of knowledge of alternative uses. Nonetheless, we can note that the expansion of coastal operations should add to net employment in the affected nations, especially where the operation is highly labor-intensive.

The expansion of small-scale operations for export earnings also should not detract from current employment levels. The adoption of this new technology should tend to cause the income distribution to be less unequal while—for a few species—contributing positively to foreign exchange earnings.

✳ *Chapter Eight*

Conclusions and Future
Research Needs

MAIN FINDINGS

Fourteen presently aquacultured species (including one plant, a type of seaweed) were selected for study. The criteria used in the selection process were evidence of existing aquaculture enterprise somewhere in the world, environmental conditions, potential consumer acceptability, adaptability to small-scale farming, and labor-intensiveness. A sample of 90 developing nations—more or less the standard textbook population—were also chosen as potential species receivers.

Species exhibiting highest and average potentials for transferability were derived from a simulation model. First, a generalized bioeconomic model was developed. Second, the parameters of the fourteen modeled enterprises were estimated. Third, the comparative production costs of all competing protein sectors were estimated. Fourth, a systems framework for deciding "recommended receivers" was devised. The decision matrices used in this systems approach were a substitution matrix that ranked protein-producing sectors by cost per unit of protein weight and cost per unit of qualitatively adjusted protein, a consumer acceptability matrix, and a technology acceptability matrix. (The details of the nature of the protein units, the model and the systems framework are presented in Chapters 4, 5 and 6 respectively).

Species and country receiver criteria for potential technological transfer were based upon alternative combinations of matrix "scores." The species and nations were divided between those of *highest* and those of *average* potential for successful transfer. The main difference between the "highest" criteria and the "average" is the raising of the average cost of aquaculture production for transfer (to the middle of the distribution of protein costs) for the *average* potential.

The aquaculture species with transfer potential can be further classified according to whether they are (1) substitutes or complements with present protein production sectors, (2) usable for both foreign exchange earnings *and* domestic consumption, or (3) usable exclusively for foreign exchange.

(1) Those species found to be substitutes or complements for present domestic protein sectors are in the order of number of countries of *highest* potential: Mullet (55 countries); Oysters (32 countries); Tilapia (30 countries); Rainbow Trout (20 countries); Milkfish (18 countries); Channel Catfish (10 countries); and Mussels (one country). Those species found to be domestic substitutes or complements with *average* potential are: Tilapia (22 countries); Yellowtail (16 countries); Walking Catfish (16 countries); Milkfish (11 countries); Channel Catfish (seven countries); Penaeus Shrimp (seven countries); Oysters (three countries); Mussels (two countries); and Macrobrachium (two countries). As indicated in Chapter 7, except for six of the potential Tilapia producers, suitable inland or coastal land appears available or substitutable for these species. Given our caveats regarding available data, we conclude that Mullet, Oysters, Tilapia, and Milkfish have the greatest potential for transfer to the coastal areas of developing nations. Mullet already is cultivated to some extent in fifteen developing nations, Tilapia in one, and Milkfish in ten. On the other side of the transferability coin, Indian Carp, True Eel, Mussels, Macrobrachium and Blue-Green Algae had very low potentials.

(2) Four species are found to have export potential in developing nations. (Tilapia is not considered because a "world price" is not available.) In some of these producing nations consumer acceptability already exists so that the species could be produced either for home or foreign consumption. Such species are Mullet in 43 nations, Rainbow Trout in 19, Milkfish in 16, and Penaeus Shrimp in five.

(3) Though rare, there are a few nations with a production potential and no domestic market for the above four species.

These are Rainbow Trout in five countries, Milkfish in three, Mullet in three, and Penaeus Shrimp in one. The lack of a domestic market is usually the result of religious taboos.

The potential technology receivers that are simply numbered in the above summary are listed by name in Chapter 6.

It is instructive to return to our list of nations from Chapter which were classified as having present famine or chronic food shortages. (1) There were *eight* nations experiencing *famine*. Of these, none would benefit from aquaculture transfer. (2) *Six* nations were experiencing *near famine*. Of these, *four* would benefit from the transfer of one species. They are Nigeria (Mullet), Syrian Arab Republic (Mullet), Bolivia (Rainbow Trout), and Tunisia (Tilapia). (3) *Fourteen* nations were suffering from chronic food shortages and all *except three* would benefit from the transferability of at least one species. They are one species each in Saudia Arabia, Angola, Uganda and Zaire, two each in Algeria, Iran and Iraq, three in Ecuador, four in El Salvador, six in Indonesia and seven in the Philippines.

In a number of critical food shortage areas of the developing world aquacultural protein can augment existing food production at their present levels of technology. Nonetheless, the most destitute nations can derive no benefits at present. The reason is found in the technology acceptability matrix of our decision making model. These nations fail to qualify for transferability because of the low level of their technological acceptance score. For these nations—assuming that our estimated scores are accurate—extensive technological assistance of a *general* nature is required. Without such assistance, these areas suffering from famine cannot help themselves through aquaculture production according to our criteria.

Because the minimum technological acceptance score was raised for our analysis of trade potential, it is hardly surprising that very few of these severe food shortage areas had an export potential in any of the selected species. Of the species found to have potential for export in the nations, one each was available to Iran (Mullet) and the Syrian Arab Republic (Mullet), two each to El Salvador, the Philippines and Indonesia (Mullet, Oysters), and three each to India (Mullet, Oysters, Penaeus Shrimp) and Ecuador (Mullet, Oysters, Rainbow Trout).

In general the availability of coastline for saltwater species is not a serious constraint on production and foreign exchange earnings from the species selected for transfer except where food production is not the highest economic use of the coastal zone. (The analysis of economic trade-offs in this zone was beyond the scope of the study.) Countries with extensive coastal areas could have substantial net additions to production, employment, and (in some cases) foreign exchange earnings in the culturing of Mullet, Milkfish, Penaeus Shrimp, Walking Catfish and possibly Rainbow Trout. A summary tabulation provides an overview of the share of countries (by area) that could benefit *in some positive* way from aquaculture. The number of nations, the number of species, and the shares of nations in each area are shown in Table 8-1.

As indicated in Chapter 3, the type of aquaculture operation selected in this study is relatively small and simple. The above conclusions therefore apply to the small, rural farmer and meet one of the major criteria outlined in Chapter 1. The transfer of aquaculture should tend to make the income distribution more equal if the emphasis is upon small-scale operation. Larger-scale production for foreign exchange earnings augmentation, of course, may result in less labor-intensive production. Because most of the operations would be coastal, the net addition to employment nonetheless should be a factor tending to make the income distribution more equal.

FUTURE RESEARCH NEEDS: RECOMMENDATIONS

Biological Research.   We offer these suggestions:

(1) In our review of much of the biological literature we found that the studies are either fragmentary or not well integrated for various potential user groups in other disciplines. We recommend that future biological research adopt a "systems approach" to analyses. That is, these studies should explain the complicated sequence of production relationships involved in culturing fish. In essence, we are calling for biological production functions showing the parametric relationships, for example, between rearing of fry, stocking of ponds or estuaries, growth rates of fish and stunting effects. This would enable economists to evaluate the economic significance of changing variables and possible impact on parameters that affect the profitability of the aquaculture enterprise.

(2) The paucity of studies on the biological aspects of polyculture should be remedied. The interaction of two or more species in culturing is a technology that is spreading rapidly.

(3) Biologists (as well as economists) should make every effort to work on an interdisciplinary basis so that priorities can be better established and programs justified.

Table 8-1.   Potential Beneficiaries of Aquaculture

Area	Number of Countries	Share of Area (%)
Africa (31 species)	23	74.2
Asia (30 species)	20	66.6
South America (10 species)	9	90.0
Central America (10 species)	10	100.0

Economic Research. The main limitations of this study derive from the lack of substantial quantities of high quality data. This lack limits the number of models and the range of tests that can be applied to the type of static simulation model developed in this study. That is, we need more and better economic data on the "typical" enterprises.

There are several priority areas. (1) Even more data are required to investigate two important economic relations: (a) The elasticity of technical substitution between factor inputs should be estimated. Since low labor costs exist in developing countries, these nations may be able to change the ratio of capital or other inputs to labor and thereby create more employment. (b) Knowledge about economies of scale would be useful. The latter investigation is extremely important since it has implications for the smaller farmer, the income distribution, and foreign exchange earnings. If significant economies of scale exist, it may be wise for the country to organize small farms into cooperatives for either export or local consumption.

(2) Demand analysis of consumer acceptance of various fishery products by country are required as these are so critical to feasibility studies.

(3) Research on regional variations of factor prices within a country should be made and location theory used to place the new aquaculture enterprise in the optimum location.

(4) Biologists, economists, sociologists and anthropologists should make some studies on the proper methods and materials needed to enable people in the developing countries—especially in Latin America and Africa—to utilize this unfamiliar technology.

EPILOGUE

In our research of the literature on aquaculture, we have observed a heavy emphasis on biological studies. Biological information is an indispensable input into creating viable economic aquaculture enterprises. However, there is a basic lack in the area of interdisciplinary work. Yet the success or failure of aquaculture critically depends on its ability to compete with alternative ways of producing protein. Hence, it is important that *extensive economic studies* be integrated with the on-going biological work. It is apparent that many biologists such as Bardach and Ryther are attempting to make economic inferences on the basis of gross productivity ratios such as pounds of fish per acre versus pounds of beef. Economics, as is biology, is more complicated than those outside the economics discipline apparently realize. Many of these complications are revealed in the present study.

The authors believe that this study has shown that aquaculture has a role to play in augmenting the world's protein supply. However, we would be very cautious in endorsing the conclusions reached by Bardach and others on the rate of growth of aquaculture production. As in the cases of agriculture and the wild fishery resources of the sea, aquaculture is not the great panacea, but only one of many ways the population/food dilemma can be mitigated. We wish that we could have laid the ghost of Malthus to rest. But he may have to be sedated with a medicine stronger than what one economic sector can provide. The "miracle" medicine most likely will be a combination of approaches that ultimately will exorcise this pervasive and restless ghost.

Annotated Biological Bibliography for Aquaculture

INTRODUCTION TO THE BIBLIOGRAPHY

The following bibliography is in two parts: (1) an annotated bibliography for the biology of aquaculture; (2) an unannotated bibliography for the economics of aquaculture. The annotated biological bibliography was prepared by a member of our research team, Randy Martin. The economics bibliography was prepared by John Vondruska of the U.S. National Marine Fisheries Service. The second bibliography does not include annotation but will be available in annotated form as John Vondruska, *Aquaculture Economics Bibliography*, NOAA Technical Report, U.S. Department of Commerce (forthcoming).

Citation gaps for particular species reflect mostly the state of the literature rather than incompleteness of the search. The authors wish to thank Randy Martin for his assistance and John Vondruska for his willingness to supply the economics bibliography.

1. CARP

Al-Hamed, M.I. *Carp Culture in the republic of Iraq (Cyprinus Carpia).* FAO Fish. Rep. 2(44):135-142. May 1967.

Fish culture in ponds was started for the first time in Iraq in 1955. The carp is the principal species cultivated; but some *Barbus* species are also reared separately or in combination with carp. Feeding experiments indicated that carp production increased four to five times by using corn, barley or cotton seed meal. About one ton of any of the mentioned feeds per donum (1 donum = 2,500 sq. m.) per season produced 350 to 450 kg. while that without feeding did not exceed 86 kg. per donum. Commercial fertilizers 13-13-5.5, N-P-K trace elements and 8-9-2, N-P-K, increased the productivity from 200 to 250 kg. of fish per donum on an average.

Alikunhi, K.H. *Synopsis of biological data on Common Carp (cyprinus carpia).* FAO Fish. synopsis, No. 31.1, 83, 1966.

Alikunhi, K.H., K.K. Sukunaran and S. Parameswaran. Induced spawning of the Chinese Grass Carp, *Ctenopharyngodon idellus* and the Silver Carp, *Phyophthalmichthys molitrix* in ponds at Cuttack, India. *Proceedings of Indo-Pacific Conference.* 10 (II), 1961.

The paper discusses secual maturation in ponds of Chinese Carp (Grass Carp, Silver Carp and Bighead). Fertilizable eggs and viable milt were obtained by intramuscular injection of pituitary extracts, homoplastic, and hoteroplastic using Indian carp as donors. Rate of hatching and subsequent survival and growth of the fry is described.

Bishai, H.M. *Experimental studies on feeding the common carp Cyprinus carpio L. in Egypt.* Bulletin of the Institute for Oceanography and Fisheries. 2:275-295, 1972.

Buck, D.H., C.F. Thoits, and C.R. Rose, *Variation in carp production in replicate ponds.* Trans. Amer. Fish. Soc. 99(1):74-79. 1970.

Variations in the production of carp, *Cyprinus carpio*, were measured in 4 consecutive seasons in 9 similar, contiguous ponds. These were believed to be the most extensive uniformity studies made of fish production in ponds. The fertility of the carp ponds varied from year to year partly due to fertilization; the means and coefficients of variation of net production in kg/ha in replicate ponds within years were, respectively; 285, 0.179; 196, 0.143, 140, 0.227; 427, 0.123. Variations in weight increments made by similar numbers of tilapia, *Tilapia mossambica* were measured over 50-day periods in 10-ft. diameter plastic pools having meticulously standardized soil substrates. These data suggest: that carrying capacity in ponds is less stable than commonly believed, that no environmental factor, such as basic fertility, maintained a dominant, continuous influence, and that in each new season production was controlled by a new assortment or combination of factors.

Chervinski, J., B. Hepher, and H. Tagari, *Studies on carp nutrition: II. Effect of a protein-rich diet on fish yields in farm ponds.* Barmidgeh (20(1):6-15. 1968.

Feeding experiments were carried out during 1966 on *Tilapia xilli*. Controls were fed sorghum, experimental fish pellets containing sorghum, wheat, corn, soybean oil meal, fish meal and fermentation products of 10 percent protein content. While the average yield from pellet-fed ponds varied among the five ponds used in the study, the average increase for two growing seasons in the five ponds was 39.9 percent higher than that from the sorghum-fed ponds.

Chiba Kenji, *Studies on the carp culture in running water pond: V oxygen supply and consumption in the fish pond and estimation of the amount of harvestable fish in a pond.* Bell Freshw. Fish. Res. Lab., Tokyo, 21(2):151-160.

Das, B.C. *Effects of micro-nutrients on the survival and growth of Indian carp fry.* FAO Fish Rep. 44(3):241-156. 1967.

Effects of micro-nutrients on the survival, growth and biochemistry of Indian carp fry from 3 to 26 days of age were investigated. Survival is enhanced significantly by treatment with vitamin B complex or yeast. Among the micro-nutrient treatments employed, yeast increases growth most. Based on these results, an equation for projecting yield after a specified period of time was formulated. Yield is expressed as a function of the initial number of 1 day old fry, the proportion surviving for a given density and treatment, and the expected weight of any fish of a given species and age group. Biochemical determinations of total protein, non-protein N, acid phosphatase and alkaline phosphatase were carried out for untreated fry and fry treated with yeast. Treatment with yeast has a statistically significant effect on total protein and non-protein nitrogen. A significant change in in total protein, acid phosphatase and alkaline phosphates occurs with age. The significant regression of weight on total protein decreases with yeast treatment. As yeast treatment decreases the total protein per g dry weight but increases the total weight, it appears to make fry efficient as weight gaining units.

Ehrlich, S. *Studies on the influence of nutria on carp growth.* Hydrobiologia 23(1/2):196-210. 1964.

Data from 9 experiments carried out in fishponds in Poland showed that the breeding of nutria is a potent means of increasing carp production. Densities of up to 55 nutria per hectare were used and the indications are that even higher concentrations should give higher production rates of carp. Increases in carp growth of over 250 percent are recorded in the experiments. The results of the experiments are analyzed statistically.

El Bolock, A.R., and W. Labib, *Carp culture in the U.A.R. (United Arab Republic).* FAO Fis. Rep. 2(44):165-174. May 1967.

The two varieties of common carp (scale carp and mirror carp) have been successfully introduced into the U.A.R. Most of the experiments and culture of carp were conducted in Barrage, Serow and Manzallah Farms. The first two farms are experimental while the last one is a commercial farm. Carp spawned with satisfactory results in cement tanks with hard bottoms. Egg collectors made of palm leaves covered with red palm fibres are used for attachment of eggs. Nursing experiments showed that 75,000 to 100,000 carp fingerlings, about 10 cm in length, can be obtained per hectare of nursing ponds in about one month. Examination of the Egyptian carp stock revealed the fact that only 17.13 percent of the examined possessed desirable body characters. After careful selection of the mothers, this ratio increased to 67.95 percent.

Such improvement was reflected in the production per hectare, which was mmuch higher in the case of the selected fish.

Fijan, N. *Problems in carp pond fertilization.* FAO Fish. Rep. 3(44):114-124. Oct. 1967.

Fertilization experiments were run on the carp ponds of a total area of more than 300 ha for four years. Different doses of calcium and phosphorous or phosphorous plus nitrogen fertilizers did not increase yield in the majority of cases. Experiments indicated that even a negative influence of fertilization on yield could exist. When the results of experiments were applied on a country-wide scale, the use of fertilizers of fish farms was lowered by about 48 percent without any negative effect on yield and feed coefficients. Chemical analyses of inflow water and water in ponds showed high concentrations of phosphorous and nitrogen and a sufficient concentration of calcium. As inflow water influenced positively the balance of elements in ponds, fertilization under such conditions seemed to be unnecessary. Supplementary feeding with cereals in densely stocked ponds is considered to be another factor which enriched the ponds with phosphorous and nitrogen. It was presumed that the analyses of factors influencing the balance of nutritive elements in ponds would give an indication whether fertilizers should be used or not.

Hepher, B. and J. Chervinski, *Studies on carp nutrition: The influence of protein-rich diets on growth.* Barridgeh 17(2):31-46. 1965.

Two years of experiments were undertaken on the nutrition of carp. Feeding with a protein-rich diet (28-30 percent crude protein) caused an increase in pond fish yield. This yield increase is apparent when the standing stock of fish per unit of pond area increases over a certain "critical" point, which is dependent on the natural food in the pond, and which is affected by factors which effect natural food production, such as climate, soil and water conditions. Below the standing stock "critical" point, the protein-rich diet has no advantage over feeding with cereal grains. It seems that in addition to protein, which becomes at a certain stage a growth limiting factor, there are also other limiting factors which are associated with the composition of protein. Protein-rich diet affected fish flesh composition also. Fish fed on this diet contained more water and protein and less fat in their flesh.

Hickling, C.F. *The artificial inducement of spawning in the Grass carp. Ctenopharyn-gadonoidallaeval.* FAO Indo-Pacific Fish Conv. Tech. Paper 5, 1966.

In this historical review and summary of spawning attempts Grass carp were most difficult to spawn, the Silver carp easiest, while the Bighead is intermediate.

Janecek, V. *Vliv slozeni krmivovych smesi na rust kapru a obsah tuku v jekich tele [The influence of the composition of feed mixtures on the growth of carp and the fat content of their bodies.]* Vzyk. Ustavu Rybarskeho a Hydrobiolog. Vodn. 7:59-92. Bibliogr. 1967. English summary.

In the course of three years (1964-66) an investigation was carried out regarding the effects of different compositions of feed mixtures and of the intensity of supplementary feeding on growth and of further economic indices, and on the simultaneous depositing of fat. The tests established in cage departments and experimental fishponds were carried out with feeds with a narrow, medium wide, and wide ratio of nutrients with normal and with an increased intensity of supplementary feeding. From the results obtained it can be seen that production was strongly influenced both by the composition and quality of feeds and by the intensity of supplementary feeding. The highest production of 3300 kg per hectare was obtained with a mixture with a wide nutrient ratio and with a content of nitrogenous substances of from 10 to 15 percent, and that both in an environment rich in natural feeds and in an environment poor in natural feeds. These results indicated a possibility of saving proteins in carp feeding, which simultaneously confirms also Lieder's (11) and Janecek's (7) findings.

Kessler, S., M. Wohlfarth, M. Lahmann, and R. Moav, *Monosex culture of carp.* Bamidgeh 13(3/4):57-60. 1961.

Growth of carp was better in fattening ponds when the sexes were separated, for this prevented wild spawning activity. Commercial sexing, easily undertaken either in winter or spring could be accomplished with only 1-4 percent error. Females grew faster than males, both in single and mixed sex populations.

Katsuzo Kuronuma and Kazuo Nakamura, *Weed Control in Farm Pond and Experiment by Stocking Grass Carp.* Proc. Indo-Pacific Fish. Coun., 7 (II):35-42, 1957.

Consideration was given to the ill effects of excess growth of aquatic plants on the productivity of fish in farm-ponds, and the methods of control currently practiced. On three farm ponds an experiment was carried out to test weed control by stocking with Grass carp, comparing the abundance of vegetation of each pond with and without stocking. The total fish production was also traced along with the Grass carp experiment. It is tentatively stated that Grass carp numbering over 30 and weighing over 35 kg if stocked for each ha of water, will curb the growth of the weed, especially submerged plants, within one year in an ordinary farm pond. It was also shown in the experiment that the stocking of Grass carp did not diminish the total fish production of the pond, though some effect was apparent according to the species among the whole fish population.

Lahav, M., S. Sarig, and M. Shilo, *The eradication of Lernaea in storage ponds of carps through destruction of the copepodidal stage by Dipterex.* Barnidgeh 16(3):87-94. 1964.

Carps infested with Lernaea were placed in aquaria containing one of four test insecticides: gammezane, malathion, Dipterex or D.D.V.P. Concentrations of grammexane and malathion of sufficient strength to destroy the parasites were toxic also to the carp. In 0.5 percent solutions of Depterex and of D.D.V.P. the fish were not harmed, nor were the adult and paupliar copopods. On the other hand, the copopodid stages of the Lernaea were killed rapidly at this concentration. Ponds with heavily infested fish were treated with the insecticides to give a concentration of 0.5 percent. The copepodids were greatly reduced on the fish after one treatment. Adult Lernaea disappeared after 3-4 treatments at weekly intervals.

Larkshmanan, M.A.V., D.S. Murty, K.K. Pillai, and S.C. Banerjee, *On a new artificial feed for carp fry.* FAO Fish. Rep. 3(44):373-387. Oct. 1967.

A series of laboratory experiments on rearing of carp fry from hatching to 15 day old fry were conducted using 19 cheap and easily available items of feed. Besides individual items, the promising items of feed were prepared into combinations of two, three, four and five and in different ratios. Survival and growth of fry were taken as the criteria for determining the relative value of the various items of food. The data were statistically analyzed in detail comparing the results of all experiments and the conclusion was drawn that a mixture of notonectics. Insects, small prawn and cow-pea in the ratio of 5:3:2 yielded most satisfactory results. Any other ratio of the three items did not show comparable results. The advantages of using the aquatic insect and prawn populations and comparative economy of the supplementary feed are discussed.

Laventer, H. and Z. Perah, *Preliminary observations on late spawning of carp.* Bamidgeh 18(2):31-36. 1966.

Wild spawning in carp fattening ponds during the spring season may be prevented by the use of late-spawned fish (spawned in July or August). Late-spawned carp fail to spawn during the first year of their life owing to a retardation in the development of their sexual organs, but grow at the same rate as normally spring-spawned fish. Late-spawned carp are particularly suited for storage for carp ponds emptied only once a year (in autumn). The late spawning season continues from July until September. Carp spawned in July seemed to have an economic advantage over those spawned in August for nursing in fattening ponds during the subsequent spring season. Carp spawned at the end of August or the beginning of September may be used for the mixed nursing population of fattening ponds early during the subsequent spring season.

Marek, M. *The effect of carp size on the food coefficient.* Bamidgeh 18(1):14-25. 1966.

The food coefficient is calculated as the ratio of total food requirement to daily weight increment of the fish. In the paper all of the data used are of a schematic nature, but they are based on an analysis of data obtained in studies of fish in ponds in Israel. It is usual in Israel to feed fish in storage ponds at a rate of 1 percent of the weight of the fish each day just in order to maintain their original weight. It is shown, because of differences of surface area to weight in fishes of different sizes, that actually the amount required amounts to 1.70 percent for fish weighing 100 g to 0.80 percent for 1000 g fish. The food coefficient is larger for fish of greater size than for smaller ones when all sizes of fish are gaining equally in weight per day, but it is smaller for large fish when the daily growth is proportional to the original weight. Hence, for 400 g fish gaining 4 g per day the food coefficient is 3.05, whereas for 1000 g fish gaining 10 g per day the coefficient is only 2.80. Similarly, food coefficients decrease for fish of a given size the more rapidly they are growing. An increase in population density of the fish, because it leads to a decreased growth rate, results in an increase of the food coefficient. On the other hand, the coefficient is smaller in ponds of greater fertility. Fish of different ages utilize differing proportions of natural and of artificial foods, and this must be taken into consideration in calculating the amount of artificial foods to be added to ponds. Adjustments must be made to prevailing pond conditions; these conditions vary seasonably as well as from pond to pond. Usually ponds are assessed either by examining the plankton and benthos content of the pond, or by growing fish in the pond for awhile on natural food only, to measure their rate of growth. The latter method is costly and it is ineffective when usual procedures are used. An improved method is described.

Meske, C. *Breeding carp for reduced number of intermuscular bones and growth of carp in aquaria.* Bamidgeh 20(4):105-119. 1968.

Methods for selection of carp for reduced number of intermuscular bones using x-rays and a television camera reacting to x-rays are described. Growth and nutrition studies, breeding experiments, and equipment used in the genetics work is also described. Sperm and egg preservation systems are suggested. No immediate results from the long-range plan have yet been obtained.

Nair, K.K. *A preliminary bibliography of the grass carp Ctenopharyngodon idella Valenciennes.* FAO Fish. Circ. 302. 1968. 15 p.

Robel, R.J. *Weight increases of carp in populations of different densities.* Trans. Amer. Fish. Soc. 91(2):235-237. 1962.

Growth rate data were collected during 1959, 1960, and 1961 from confined carp (*Cyprinus carpio*) populations approximating 200, 400, and 600 lb./acre. The weights of the populations were recorded when they were introduced into enclosures early in the summer and again when they were removed in the fall. All the carp population showed increases in weight during the confinement periods, with the lower population levels exhibiting the greatest increases. The causes of the relationship between population density and growth rate were not determined, but the food supply appeared not to be depleted in any of the pens.

Sarib, S., M. Lahav and M. Shilo, *Control of Dactylogyrus vastator on carp fingerlings with Dipterex*. Bamidgeh 17(2):47-52. June 1965.

Schaperclaus, W. *Etiology of infectious carp dropsy*. Ann. N.Y. Acad. Sci. 126(1):587-597. 1965.

Causes of dropsy in carp were examined with results that point to *A. punctata* as the main etiologic agent of infectious dropsy.

Shimma, Y. *An experiment of artificial feed coated by myristic acid for carp larvae*. Bull. Freshwater Fish. Res. Lab. (Tokyo), 17(2):107-112. 1967.

Because of lower growing rate of larvae if compared with results by natural feed stuffs, more studies on coating materials, feed components, need to be undertaken. This study uses skim milk and dried egg powder fed to the larvae of common carp.

Singh, S.B., K.K. Sukumaran, K.K. Pillai, and P.C. Chakrabarti, *Observations on efficacy of Grass carp, Ctenopharyngodon idella in controlling and utilizing aquatic weeds in ponds in India*. FAO Indo-Pacific Fish Coun. Tech. Paper 20. 1966.

Experimental study could control thick infestations of submerged and floating weeds, among others. Information on effective stocking given.

Soller, M., Y. Shchori, R. Moav, G. Wohlfarth, and M. Lahman, *Carp growth in brackish water*. Bamidgeh 17(1):16-23. 1965.

The potential importance of fish culture in brackish water is outlined. Two series of experiments aimed at investigating genetic factors influencing carp's growth in brackish water are described.

Steffens, N., M.L. Albrecht, D. Barthelmes, H. Kulow, U. Lieder, H.U. Menzel, and G. Predel, *Results of cage rearing of carps in cool water from power stations*. Z. Fisch. Hilfswiss. 17:47-77. 18 ref. 1969.

Experiments were performed on carp rearing in warm water from the cooling water outlet of powerplants. The results of physiological investigations on organ weight composition, and fat tissue, and the economics of caged carp rearing are discussed.

Stevenson, J.H. *Observations on grass carp in Arkansas*. Prog. Fish-Cult. 27(4):203-206. 1965.

Early growth of the grass carp at the Fish Farming Experimental Station, Stuttgart, Arkansas, compared favorably with that reported in semitropical countries. The average weight at 18 months was 1,816 g; the length, 50 cm. The fish were given a supplemental ration of commercial fish pellets and cut grass. Observations of the feeding habits indicate that this carp may not be a strict herbivore; it is recommended that a thorough study be made before the fish is released in natural waters.

Stott, B., and L.O. Orr. *Estimating the amount of aquatic weed control consumed by grass carp.* Prog. Fish-Cult. 32(1):51-54. 1970.

Sukhoverkhov, F.M. *The effect of cobalt, vitamins, tissue preparations and antibiotics on carp production.* FAO Fish. Rep. 44(3):400-407. 1967.

Antibiotics are evidently concerned with protein metabolism and can make up to some extent for lack of animal proteins in the rations. On the basis of these experiments the addition of these growth-stimulating substances to feed mixtures for carp is recommended.

Szumiec, M. *Thermal characteristics of carp ponds.* FAO Fish. Rep. 44(3):265-273. 1967.

Observations on temperature variations were carried out in spawning, nursery and rearing ponds. In all ponds distinct differences exist in time-space temperature distribution. These differences depend on the size of the pond, depth being the decisive factor.

Tan Lee Wah, *Carp Culture in Singapore: A Case Study.* J. Trop Geogr. 35:67-74.

Example of how a mangrove swamp area can profitably be connected to a fish pond for the production of carp.

Tashpulatov, E.A. *Cultivation of carp in rice paddies of Western Fergana and the effect of fish on rice yield.* UZB Biol ZH 15(6):62-63, 1971. (in Russian.)

Carp grown in rice paddies increases yearly rice yield. Used 500-550 fish/ha.

Tesarcik, J. *Vyskyt branchionekrozy (branchionecrosis cyprinorum) u kapri nasady v severomoravskem kraji [The occurrence of branchionecrosis in a carp population in northern Moravia].* Bul. VUR Vodnany 4:3-6. 1969. English summary.

This paper contains a description of the occurrence and course of branchionecrosis in a carp population in northern Moravia.

Vaas, K.F. and M. Sachlan, *Cultivation of common carp in running water in West Java.* FAO Indo-Pacific Fish Council Tech. Paper 3, 1955.

Common carp is described in Indonesia. Use of bamboo cages, along with water analysis, gut contents and growth rates are described.

Vaas, A. *Experiments on Different Stocking Rates of the Common Carp.* Proc. Indo-Pacific Fish Coun. 7(II):13-34, 1957.

A series of experiments on stocking rate was carried out in the hatchery of the Laboratory for Inland Fisheries, Bogor, Indonesia. The individual growth and the total yield at different stocking rates were determined. Results conclude that at the optimal stocking rate the minerals are most rapidly incorporated by the carp into the flesh of their body; in this case they are no longer available for phytoplankton.

Vaas, A., *On the Use of Marygold as Green Manure in Indonesian Carp Ponds*. Proc. Indo-Pacific Fish. Coun. 7(II):1-11, 1957.

A series of experiments with *Tithonia dive sifolia* Gray (Marygold) was carried out in order to increase the yield of carp. It was concluded from investigations concerning yield, the biota of the ponds and the gut contents of the fish, that maximal increase in yield is obtained by manuring in the inflowing water, in ponds stocked with fry of the Common carp together with large male Tilapia.

_____. *Studies on Food and Feeding Habits of the Common Carp in Indonesian Pond.* Proc. Indo-Pacific Coun. 7(II), 1957.

Verma, Mahendra N. *Breeding of major Indian carps in a fountain pond.* Prog. Fish. Cult. 32(4):222-223. 1970.

Major Indian carp do not breed in confined waters. 91 percent of spawn cultured is obtained from rivers, 9 percent from pituitary injection and bund breeding. Natural spawning was observed in a pond when some were injected. Three possibilities account for this. 1) Social status and behavior. 2) Gonadal secretion of injected fish. 3) "turmoil and confusion" and gonadal secretions.

Wohlfarth, G., R. Moav, and M. Lahman, *Genetic improvement of carp. III. Progeny tests for differences of growth rate, 1959-1960.* Bamidgeh 13(2):40-54. 1961.

Three seasonal progeny testing experiments were carried through during the summer and fall of 1959 and in the spring of 1960. The individual progenies showed much the same relative growth rates in all three seasons. Interactions between different progenies were observed as was crowding conditions.

Woynarovich, E. *Hatching of carp eggs in "Zuger" glasses and breeding of carp larvae until an age of 10 days.* Bamidgeh 14(2):32-46. 1962.

A method of artificial breeding of carp is described. The eggs and sperm are stripped from the fish and mixed in a little water to ensure fertilization. The eggs, however, are sticky and will not develop naturally when clumped together. The sticky substance is removed using NaCl and thereafter are placed in "Zuger" glasses and allowed to develop.

Yashouv, A., E. Berner-Samsonov, and K. Reich. *Forced spawning of silver carp.* Bamidgeh 22(1):3-8. 1970.

Silver carp were grown in fishponds in Israel from fingerling size. Normal spawning in Formosa occurs at 6-8 years. Experimental fish were injected with pituitary extract at an age of 2 years. Successful spawning and fertilization of eggs was attained artificially, and complete development of the embryos in 5 out of 15 cases.

2. CATFISH

Allen, K.O., and J.W. Avault, Jr. *Effects of brackish water on ichthyophthiriasis of channel catfish.* Prog. Fish-Cult. 32(4):227-230. 1970.

Allen, Kenneth O. and James W. Avault, Jr. *Notes on the relative salinity tolerance of channel and blue catfish.* Prog. Fish-Cult. 33(3):135-137, 1971.

Tolerance to salinity: white catfish, blue catfish, channel catfish. Salinity used was 14 ppt.

Andrews, J.W. *Nutritional aspects of intensive catfish production.* Publ. Skidaway Inst. Oceanogr. 1:68-77. Bibliogr. 1970.

It is pointed out that in intensive catfish culture all nutrients have to be provided in contrast to pond culture in which some are found in the environment. The basal diet worked out by SIO scientists is given with the results of experiments on growth and energy requirements.

Andrews, J.W., L.H. Knight, and T. Murai. *Temperature requirements for high density rearing of channel catfish from fingerling to market size.* Prog. Fish-Cult. 34(4):240-241.

Optimal temperatures for growth were 28° and 30°C. Food conversion—1.5 or less.

Bonn, E.W. and B.J. Follis. *Effects of hydrogen sulfide on channel catfish (Ictalurus punctatus).* Trans. Amer. Fish. Soc. 96(1):31-36. 1967.

The natural production of sulfides is responsible for poor channel catfish production in many acid lakes in northeast Texas. Channel catfish populations can be maintained by continued stocking of adult fish or by raising the pH with agricultural limestone, which in turn lowers the toxic un-ionized hydrogen sulfide.

Brown, E.E. and J.L. Chesness. *Hypothetical costs for earthen raceway culture of channel catfish.* Publ. Skidaway Inst. Oceanog. 1:33-36. 1970.

The author gives the hypothetical costs for 25 earthen raceway segments using different stocking rates. Annual costs for rearing fish varies with stocking rates and hypothetical costs are given for each rate.

Brown, E.E., M.G. LaPlante, and L.H. Covey. *A synopsis of catfish farming.* Ga. Agric. Exp. Sta. Bull. 69. Sept., 1969. Bibliogr. 50 p.

The purpose of this study was to compile and present the contemporary research on channel catfish farming, and develop an analysis of the costs and returns of channel catfish farming in Georgia. To produce channel catfish on a commercial basis the fish farmer must know the fundamentals of spawning, disease control, nutritional requirements of fish, harvesting, pond construction, and be aware of potential markets for his fish. These areas were given particular attention in this study. Much of the data for this study was obtained from individuals and institutions currently conducting research on channel catfish farming. The data concerning Georgia producers was obtained through personal interviews with the commercial producers in the state.

Bryan, R.D. and K.O. Allen. *Pond culture of channel catfish fingerlings.* Progr. Fish-Cult. 31(1):38-43. 1969.

Methods of producing channel catfish (*Ictalurus punctatus*) fingerlings at the Tupelo National Fish Hatchery from 1959 to 1966 are described.

Crance, J.H. and L.G. McBay. *Results of tests with channel catfish in Alabama ponds.* Progr. Fish-Cult. 28(4):193-200. 1966.

The average catch of channel catfish (*Ictalurus punctatus*) was 2.2 and 1.65 per acre during the 1st and 2nd years of fishing respectively, in ponds containing established populations of largemouth bass (*Micropterus salmoides*) and bream (a combination of bluegill, *Lepomis macrochirus*, and red-ear sunfish, *L. microlophus*) and stocked with 100 or 200 channel catfish per acre.

Crawford, B. and A. Hulsey. *Effects of M.S. 222 on the spawning of channel catfish.* Progr. Fish-Cult. 25(4):214-215. 1963.

Hatcherymen must sometimes handle large numbers of large channel catfish during sexing and pairing procedures. These fish can inflict dangerous wounds, and it would be useful to know whether the anesthetization of the fish with tricaine methanesulfonate (M.S. 222) would effect their spawning. Tests were made with several catfish, and the authors conclude that M.S. 222 does not adversely affect the success of the spawn or the viability of the fry when used during sexing or pairing.

Deyoe, C.W., O.W. Tiemeier, and C. Suppes. *Effects of protein, amino acid levels and feeding methods on growth of fingerling channel catfish.* Prog. Fish-Cult. 30(4):187-195. 1968.

Studies were made to determine the effects of protein levels, vitamin additions, added amino acids and different feeding systems on channel catfish, *Ictalurus punctatus*, fingerlings.

Dupree, H.K., O.L. Green, and K.E. Sneed. *The growth and survival of fingerling channel catfish fed complete and incomplete feeds in ponds and troughs.* Progr. Fish-Cult. 32(2):85-92. 1970.

Three feeds, two dietarily complete and one dietarily incomplete, were fed to fingerling channel catfish (*Ictalurus punctatus*) in ponds and in troughs. In ponds, no difference in gain, feed conversion, or survival could be attributed to any feed. In troughs, more fish died when fed one complete feed and the incomplete feed than when fed another complete feed, but there was no weight gain difference on either complete feed. Weight gain of the fish on the incomplete feed was significantly lower than weight gain of the fish on either complete feed. After fortification of the incomplete feed with vitamins or beef liver, deaths were about the same as those of fish fed the best complete feed. Gain after beef liver supplementation approached that of fish on the best complete feed. Nutrient quality of feeds appears to be of little importance if natural feeds are available in amounts sufficient to balance the diet.

Geibel, G.E. and P.J. Murray. *Channel catfish culture in California.* Prog. Fish-Cult. 23(3):99-105. 1961.

Culture of channel catfish has been successfully accomplished during 1958-60 in a California State hatchery.

Gray, D.L. *The biology of channel catfish production.* Arkansas Agric. Ext. Serv. Circ. 535. May 1969. 19 p.

The purpose of this study was to compile and present recent work on channel catfish farming with emphasis on biological aspects. Pond construction, techniques of fingerling production, spawning methods, stocking and feeding rates, feeds, control of fish diseases are included.

Greenfield, J.E. *Some economic characteristics of pond-raised enterprises.* Proc. Comm. Fish Farming Conf. (Ga. Univ.) 1:67-76. 1969.

Although Delta catfish farming economics cannot be directly transposed to other areas under different conditions, they may be helpful in bracketing the importance, variability, and potential for improvement in some of the major cost factors. The discussion of some of the market forces at work in the industry may indicate the nature of future competition and bring less efficient production ventures into a shorter, more realistic time focus. Many of the related business opportunities such as fingerling production and the production of brood stock were beyond the scope of this analysis but may offer special, higher-risk, but potentially more profitable business opportunities.

Hastings, W. and H.K. Dupree. *Formula feeds for channel catfish.* Progr. Fish-Cult. 31(4):187-196. 1969.

Rations containing fish meal resulted in better growth than those containing nonsupplemented soybean meal as a substitute. Though lysine, methionine, or dried fish solubles improved growth of fish fed soybean meal rations, none of these supplements were economical at the levels used. Vitamin additives appear essential for growth and survival of catfish in aquariums and are economical for fish in ponds. Feeds containing large amounts of fiber provide satisfactory growth when physical properties are acceptable. Milo, substituted for high-fat bran, provides adequate growth, but lowers feed conversion.

Heaton, E.K., T.S. Boggess, and D.R. Landes. *Some evaluations of tank cultured channel catfish.* Pub. Skidaway Inst. Oceanog. 1:41-49. 1970.

This paper gives the results of evaluation studies on tank cultured catfish. The tests were of two types, sensory and physico-chemical. Sensory data collected was of appearance, aroma, color, texture and flavor for both breaded and unbreaded fillets. Physico-chemical tests consisted of shear press measurements, protein changes, fatty acid changes and dress-out weights.

Imam, A.E., H.M. Roushdy and A. Philisteen. *Feeding of catfish Clarias layera in experimental ponds.* Bull. Inst. Oceanogr. Fish. 1:205-222. 1970.

Tried various plant and animal artificial foods. The three types used were 1) minced fresh forage fish meat, 2) rice bran, and 3) fresh vegetables. Number 1 worked best because of better growth rates and better health conditions.

Matsuda, Y. *Disease and parasite control in intensive catfish culture.* Pub. Skidaway Inst. Oceanog. 1:78-79. 1970.

While disease problems tend to increase with stocking density, tank or raceway culture systems offer many advantages over pond culture in regard to disease and parasite control. Advantages are: the fish are easily viewed, therefore disease and parasite problems can be detected, the producer can afford to use expensive drugs in a high density population, preventive techniques can be easily applied to a high density system, and water turnover flushes disease organisms from the tank. The high cost of clearing drugs with FDA is discussed and the drugs most widely used in Japan are listed.

Morris, A.G. *Production of channel catfish to creel size.* Progr. Fish-Cult. 29(2):84-86. 1967.

Murphy, J.P. and R.I. Lipper. *Experiments work toward early catfish spawning.* Amer. Fish Farmer 1:16-20. Sept. 1970.

Experiments with warm-water ponds were conducted in Kansas to induce channel catfish spawning. The purpose was to advance spawning date by several weeks in an effort to produce marketable fish in two seasons rather than three.

Prather, E.E. *Over a ton of channel catfish per acre.* Auburn, Ala., Alabama Agricultural Experiment Station, 1968. Reprinted from Highlights of Agricultural Research 15(1):4.

This paper reviews experiments in catfish culture being carried out by Auburn University Agricultural Experiment Station. Pond requirements including water depth, water movement and fertilization are included.

Regier, H.A. *Ecology and management of channel catfish in farm ponds in New York.* N.Y. Fish Game J. 10(2):170-185. July 1963.

Experiments with channel catfish were conducted in farm ponds in central New York. This species does not appear to hold much promise for general stocking in New York farm ponds because it is difficult to produce cheaply and in large numbers for stocking purposes; survival and reproduction of stocked fish were poor in most project introductions; and the unpredictability of survival and of reproduction implies that harvest quotas would be unpredictable.

Sandoz, O. *Experimental feeding and growth of channel catfish in Oklahoma hatcheries (Ictalurus punctatus).* Proc. Okla. Acad. Sci. 47:414-421. 1969.

During the winter of 1965-66, seven hatchery ponds were stocked with advanced fingerling channel catfish and reared to subadult size for stocking in various Department of Wildlife Conservation Lakes.

Shell, E.W. *Effects of changed diets on the growth of channel catfish.* Trans. Amer. Fish Soc. 92(4):432-434. 1963.

Troughs of channel catfish that had been fed purified diets containing different levels of casein as the only protein source were placed on a multiple protein diet containing a single level of protein. Growth rate of the fish while on the single protein diet was directly related to the amount of protein in the diet. Growth rate on the multiple protein diet during the first 2-week period was inversely related to the growth rate of the same fish while being fed the single protein diet. These results suggest that, after receiving the same diet for a period of time, a short period may be required for readjusting the metabolic machinery for the new protein. The amount of time needed for this readjustment could conceivably be related to the rate of growth on the old diet, or to the total amount of protein assimilated while the fish were on the old diet.

Shrable, J.B., O.W. Tiemeier, and C.W. Deyoe. *Effects of temperature on rate of digestion by channel catfish.* Progr. Fish-Cult. 31(3):131-138. July 1969.

Evidence is presented describing the effects of different water temperatures on rates of digestion by channel catfish in an effort to determine the optimum water temperature for these rates.

Simco, B.A. and F.B. Cross. *Factors affecting growth and production of channel catfish, Ictalurus punctatus.* Univ. Kans. Publ. Mus. Natur. Hist. 17(4):191-256. 1966.

Production of channel catfish under various experimental conditions was carried on in 1/10-acre earthen ponds and other facilities of the State Biological Survey at Lawrence, mostly from 1960 through 1964.

Thompson, K.W. and W.M. Lewis. *Common catfish diseases and parasites.* Amer. Fish Farmer 1:18-20. Nov. 1970.

The etiology and control of diseases and parasites are of interest to fish farmers and biologists alike. The authors discuss parasitic infection of catfish by *Trichodina, Ichthyophthirus, myrosporidians, Scyphidia,* and *Dactylogyrus.* Bacterial infection of the cut, and columnares (caused by *Chrondrococcus*) are also described. These commonly encountered pathogens continue to be a major cause of financial loss in catfish farming.

Snow, J.R. *A comparison of rearing methods for channel catfish fingerlings.* Progr. Fish-Cult. 24(3):112-118. 1962.

Five methods of rearing channel catfish fry to a minimum total length of 3 inches were compared.

Stickney, R. *Intensive catfish fingerling production.* Publ. Skidaway Inst. Oceanogr. 1:50-61. Bibliogr. 1970.

The author discussed methods of spawning catfish in the pond environment. Nesting sites and spawning pens with modifications are described. Experiments now being carried out at SIO to reduce the time period from egg stage to marketable fish are discussed.

Stickney, Robert R., T. Murai, and G.O. Gibbons. *Rearing channel catfish fingerlings under intensive culture conditions.* Prog. Fish-Cult. 34(2):100-102. 1972.

Stocking density should be one so that when the fish are at an individual weight of 1/2 oz. the density is 1 lb/cf (i.e., 32 fish per cu. foot water). This yields greatest growth rates without stunting.

Tiemeier, O.W. *Increasing size of fingerling channel catfish by supplemental feeding.* Trans. Kans. Acad. Sci. 65(2):144-153. 1962.

Tiemeier, O.W. *Growth obtained by stocking various size combinations of channel catfish and efficiencies of utilizing pelleted feed.* Southwest. Nat. 13(2):167-174. 1968.

When identical biomasses but different numbers of large and small age group II channel catfish were stocked in 2 polyethylene-lined ponds and fed pellets during spring and summer months, individual large fish grew consistently more than small fish but total gains of the more numerous small fish were greater.

Tiemeier, O.W., C.W. Deyoe, and S. Wearden. *Experiments with supplemental feeds for channel catfish.* FAO Fish. Rep. 44(3):388-399. 1967.

Fingerling channel catfish were fed with diets of dry feed in 12 polyethylene lined earthen ponds.

Tiemeier, O.W. and C.W. Deyoe. *Production of channel catfish.* Bull. Kans. State Univ. Agric. Exp. Sta. 508. Manhattan, 1967. 23 p.

The author has compiled information on all possible aspects of catfish culture. Construction of ponds, water supply soil characteristics, stocking, hatching, feeds, feeding devices, rates of growth, fertilizing of ponds, parasites and diseases are discussed and described. Summary and recommendations are included.

Washburn, K.W. *Potential for genetic improvement in growth and disease resistance of catfish.* Pub. Skidaway Inst. Oceanogr. 1:62-67. 1970.

The author points out the two major factors that will set the limits on growth rate in catfish: environmental factors and genetic factors. While environmental factors such as management, nutritional and disease aspects can provide the optimum environment for growth, genetic factors for growth will set the upper limits on how much improvement can be made by adjustment of the environment. Plans for studying the feasibility of selecting lines for improvement in growth rate and disease resistance and increased growth potential at lower water temperatures are described.

White, J.T. *Louisiana catfish harvest.* Amer. Fish Farmer 1:10-12. July 1970.

A demonstration of fish harvesting techniques and methods for handling and storing fish was conducted at the Pennzoil Farm in cooperation with the Bureau of Commercial Fisheries. Mechanized haul seining of the pond was accomplished with a truck-mounted winch and a block and rigging attached to an anchor vehicle. In order to complete the harvest the equipment was shifted during the seining operation. Two tandem live-cars, attached by a funnel-shaped net to the seine, were successfully used for storage and loading of fish. Loading of fish by utilization of a vortex pump was unsuccessful. The design of production ponds at the Pennzoil facility are described. This catfish farming operation has been established as a pilot project to study profitable utilization of land and water resources on property acquired for mineral development.

3. TILAPIA

Allanson, B.R., and R.G. Noble, *The tolerance of Tilapia mossambica (Peters) to high temperature.* Trans. Amer. Fish. Soc. 93(4):323-332.

The median lethal temperatures of *T. mossambica* at different acclimation temperatures were estimated and the ultimate upper lethal temperature was found to be between 38.2°C and 38.25°C. The effects of acclimation temp. and period of acclimation upon resistance were also investigated. An average acclimation rate of 1° per 150 minutes was estimated for fish transferred directly from 25°C to 30°C whereas it was observed that fish transferred from 5° to 15°C were not fully acclimated even after 20 days. These data confirm that tilapia is a thermophilic species.

Chen, Tung-Pai. *The culture of Tilapia in rice paddies in Taiwan.* Fish. Ser. #2, Chinese-American Joint Commission on Rural Reconstruction, 1953.

Chervinski, J. *Growth of Tilapia aurea in brackish water.* Bamidgeh 18(3/4):81-83. 1966.

There was no significant difference, according to the t-test between the growth rate of *Tilapia aurea* in fresh water earthen ponds and in ponds containing 6°/oo and 10°/oo crude NaCl. There was significantly greater mortality however, in the brackish ponds.

_____. *Study of the growth of Tilapia galilaea (Artedi) in various saline concentrations.* Bamidgeh 13(3/4):71-74. 1961.

Tilapia were grown in freshwater and in sea water diluted 1:1 and 3:1 with freshwater. Males grew equally in all three concentrations. Females grew as well in 25 percent sea water as in freshwater.

Chervinski, J. and A. Yashouv, *Preliminary experiments on the growth of Tilapia aurea, Steindachaer (Pisces, Cichlida) in Seawater ponds.* Bamidgeh 23(4):125-129. 1971.

Tilapia were reared in seawater ponds whose salinity ranged from 36.57 percent to 44.61 percent without loss in growth. Spawning may occur.

Chimits, P. *Tilapia and its culture: A preliminary bibliography.* FAO Fish. Bull. 8(1):1-33, 1955.

_____. *The Tilapias and their culture. A second review and bibliography.* FAO Fish. Bull. 10(1):1-24, 1957.

Dadzie, S. *Laboratory experiment on the fecundity and frequency of spawning in Tilapia aurea.* Bamidgeh 22(1):14-16, 1970.

Under tank conditions, Tilapia spawned three times per year. Injection with hormones decreased the time between spawnings, but did not increase the number of spawnings per year. Incubation in the mouth of the female lasted for 8-10 days at 29°C.

_____. *Preliminary report on induced spawning of Tilapia aurea.* Bamidgeh 22(1):9-13. 1970.

Extract of carp pituitary, luteinizing hormone, and human chorionic gonadotrophin were injected singly or in combination into mature Tilapia males and females. Human chorionic gonadotrophin was more effective alone than either of the other hormones, but the best results were obtained by using human c.g. in combination with extract of carp pituitary.

Dendy, J.S., J. Varikul, K. Sumawidjaja, and M. Potaros, *Production of Tilapia mossambic Peters, plankton and benthos as parameters for evaluating nitrogen in pond fertilizers.* FAO Fish. Rep. 44(3):226-240. 1967.

Twelve 0.1 ha earth ponds were used for this study. Four ponds received no fertilization, four received non-nitrogenous fertilization, and four received nitrogenous fertilization, the same fertilizer treatments which had been applied for the previous five years. Experiment went from July to October.

Increases in weights of fish per kg were determined upon draining of ponds. Unfertilized ponds produced an average of 242.6 kg/ha. The minimum and maximum rates of stocking resulted in production of 225.9 and 284.7 kg/ha, respectively. The two fertilizer treatments resulted in productions that were essentially alike. In fertilized ponds the average productions with increase in rates of stocking were 342.0, 695.3, and 836.7 kg/ha, respectively. The study results indicated increase in production with increase in rates of stocking and from 63 to 199 percent greater production in fertilized ponds. However, the elimination of N from fertilizer for six years did not result in a decline in fish production when water temperature remained above 22°C.

Ishac, M.N. and A.M. Dollar, *Studies on manganese uptake in Tilapia mossambica and Salmo gairdnerii. I. Growth and survival of Tilapia mossambica in response to manganese.* Hydrobiologia 31(3-4):572-584. 1968.

Fingerlings, five weeks old, were reared in media either free of Mn or with small quantities of Mn supplied. Diets were either low in manganese or small quantities were added. Feeding tests lasted 10 weeks and best growth was observed when the element was supplied in both water and diet. Poor growth, reduced food consumption, loss of equilibrium and increased mortality were observed when Mn was deficient in either water or food. A daily requirement is calculated.

Kelly, Hugh David. *Preliminary studies on Tilapia Mossambica Peters relative to experimental pond culture.* Unpubl. Theses, Auburn University. 1955.

Kirk, R.G. *A review of recent developments in Tilapia culture, with special reference to fish farming in the heated effluents of power stations.* Aquaculture 1(1):45-60. Lotan, R.

Lotan, R. *Adaptability of Tilapia nilotica to various saline conditions.* Bamidgeh 12(4):96-100. 1960.

In laboratory experiments, *Tilapia nilotica* would withstand sudden changes from freshwater to 60 percent sea water, but died with hemorrhaging after sudden transferral to 70 percent sea water. When the salinity of the water was increased gradually, the majority of the fish continued to live as long as nine days in 148 percent sea water.

Mires, D. *Mixed culture of Tilapia with carp and gray mullet in Ein Hamifratz fish ponds.* Bamidgeh 21(1):25-32. 1969.

The system of growing male hybrid tilapias in their second year of life as additional fish in carp ponds together with mullet nearly solves the local problem of the year-round marketing of tilapia. The size of the male tilapia was very satisfactory, especially considering that no additional feed was added for them. Except for one pond (pond 51) no considerable amount of tilapia fry was found; even in that pond, the large amount of tilapia fry did not affect the growth of the larger fish.

Pruginin, J. *Preliminary report on the culture of Tilapia nilotica as a supplementary fish in carp ponds in Israel in 1960 and 1961.* Bamidgeh 14(1):16-18. 1962.

The following problems are discussed: reproduction, effect of the length of the growing period on the yield, rate of survival, yields, effect on the growth of carp, extra labor required, and marketing. The conclusions are based upon two years of experience in the commercial culture of Tilapia in carp ponds.

Semakula, S.N. and Makoro, J.T. *The culture of Tilapia species in Uganda.* FAO Fish. Rep. 2(44):161-164. May 1967.

Several species of Tilapia have been raised in ponds in Uganda since 1953. Because of their prolific breeding habits, the ponds soon become overcrowded. The size of fish produced is unacceptable to the consumers and the total yield from ponds is low. Little success has been obtained with predators to control the populations. Mono-sex culture through hybridization seems to be the answer to this over-breeding and low yield problem.

Shell, E.W. *Relationship between rate of feeding, rate of growth and rate of conversion in feeding trials with two species of Tilapia, Tilapia mossambica Peters and Tilapia nilotica Linnaeus.* FAO Fish. Rep. 3(44):411-415. Oct. 1967.

An experiment was conducted to determine the relationship between feeding rate, growth rate and conversion in two species of tilapia. In Tilapia mossambica Peters growth rate increased fourfold when the feeding rate was increased from 1 to 2 percent. There was little increase in growth rate when the feeding rate was increased with increases in feeding rate from 1 to 4 percent. The best conversion rate was obtained at the 2 percent feeding rate with T. mossambica and at the 1 percent feeding rate with T. nilotica. Most economical rates of feeding were lower than the rates giving maximum growth.

Silliman, R.P. *Effect of crowding on relation between exploitation and yield in Tilapia macrocephala.* Fish. Bull. U.S. Dept. of Commerce, 70(3):693-698, 1972.

Equal populations of Tilapia were set up in different areas of tanks and exploited at between 10 percent and 25 percent bimonthly. Results are given, and rates of food conversion discussed.

Swingle, H.S. *Comparative evaluation of two tilapias as pond fishes in Alabama.* Trans. Amer. Fish. Soc. 89(2):142-148. 1960.

Uchida, R.N. and J.E. King, *Tank culture of tilapia.* Bull. U.S. Fish and Wildlife Serv. 199(62):21-52. 1962.

This study evaluated the feasibility of producing bait-size tilapia by the tank culture method.

Yashouv, A. *Mixed fish culture in ponds and the role of Tilapia in it.* Bamidgeh 21(3):75-92. 1969.

Experiments were undertaken in ponds at the Fish Culture Research Station with mixed cultures of carp, mullet and silver carp.

4. MULLET

Scientists control reproduction of mullet. Com. Fish. Rev. 34(5-6):8. 1972.

Controlled reproduction of mullet (inc. spawning out of season) was done at the Hawaiian Oceanic Institute of Waimanalo. Used temperature and light controls and the females were injected with hormones. Upwelling system used to prevent larvae settling in critical first three-day period.

Hashimoto, S. *Chemical fertilization of mullet ponds.* Hawaii Acad. Sci. Proc. 29. 1954.

The use of chemical fertilizers to increase the production of fish in freshwater ponds is a common practice, but there is a dearth of corresponding information on fertilization of salt-water ponds. The results of three years of commercial fertilizer application to a mullet pond in Oahu are reported.

Mires, D. *Preliminary observations on the effects of salinity and temperature of water changes on Mugil capito fry.* Bamidgeh 22(1):19-24. 1970.

Fry were placed in salinities ranging from that of sea water to tap water with no ill effects even when they were transferred directly from sea water to freshwater and vice versa. The experiments were designed to test methods used by workers catching fry and transferring them to fishponds.

Nash, C.E., C.M. Kuo and S.C. McConnel, *Operational procedures for rearing larvae of the grey mullet (Mugil cephalua L.).* Aquaculture 3(1):15-24. 1974.

Induced spawning with gonadotropin.

Sarojini, K.K., *Observations on the Occurrence, Collection, Acclimitisation, Transport and Survival of Mullet Seed in West Bengal (India).* Proc. Indo-Pacific Coun. 7(II). 1957.

Fry of several species of Grey mullets are available in coastal and estuarine areas in Bengal. The season of greatest occurrence of the fry are listed and a key to a detailed description of the fry are given. Adaptability of mullet fry to differing salinities is discussed and experiments in the transfer of fry from salt to freshwater are described. It is concluded that such acclimatization is a simple problem and does not require a long conditioning period.

Yang, Won Tack, and Ul Bae Kinn, *A Preliminary report on the artificial culture of grey mullet in Korea.* Proc. Indo-Pacific Fisheries Council, Tech. Papers, 9th Session, Sect. II. pp. 62-70. 1962.

Hydrology and topography of grey mullet in Yungsan spawning area (estuary); techniques used to obtain and hatch *Mugil* eggs are described.

Yashouv, A. *Breeding and growth of grey mullet (mugil cephalus L.).* Bamidgeh 18(1):3-13. 1966.

The growth of grey mullet was studied in monoculture and in mixed culture with carp and Tilapia in experimental ponds. The rate of growth decreased with size of the experimental mullet.

————. *Preliminary report on induced spawning of Mugil cephalus reared in captivity in freshwater ponds.* Bamidgeh 21(1):19-24. 1969.

M. cephalus fry were gathered from brackish water in the estuary of Dalia Stream and grown to adulthood in freshwater ponds. Under these conditions the fish reach maturity, but are inhibited from spawning. Artificial spawning was induced by injections of extracts of carp pituitaries. In other experiments the urine of pregnant women and "pure LH gluteinizing hormone" were successfully utilized.

Yashouv, A. *Efficiency of mullet growth in fishponds.* Bamidgeh 24(1):12-25. 1972.

Mullet grown in monoculture and in polyculture were tested. They did better in polyculture. Ponds were fertilized with superphosphate and ammonium sulfate.

Yashouv, A. and others, *Preliminary report on induced spawning of M. cephalus reared in captivity in freshwater ponds.* Bamidgeh 21(1):19-24. March 1969.

Previous studies showed that *M. cephalus* would mature in fresh and bracking water but spawning was inhibited. In the present experiments spawning was induced by treatment with carp pituitary extracts in the freshwater environment. Fertilization was achieved and prelarvae survived four days.

5. MILKFISH

Djaingsastro, A. *Paddy-cum-milkfish culture, a creation from the lowlands of the Solo River.* Proc. Indo-Pacific Fish. Coun. 7(sect.2/3):163-166. 1959.

An introduction of paddy-cum-milkfish culture, a creation from the lowlands of the Solo River, is presented. This method has been started and developed by the farmers of that region, in their efforts to increase their income, as their paddy fields do not give any possibility for improvement. These paddy-cum-milkfish ponds are created by digging circular moats inside the paddy fields and are irrigated by a combination of rain and water from the irrigation canals scooped into them by manual labor. It is evident that from the technical, economic, and social point of view, the prospects of this kind of fish culture are favorable.

Djajadiredja, R.R. and Ruchiat Amidjaja, *Some Observations on Chanos Culture in Fresh Water.* Proc. Indo-Pacific Fish. Coun., 8(II):9-18, 1958.

This is a study in the Chanos culture known as "paddy-cum-milkfish" culture.

Rabanal, H.R. *The Elevation of a Swampland Based on the Tidal Datum and its Importance in Selecting Sites for Chanos Fishponds Projects.* Proc. Indo-Pacific Coun. ID(II) 1961.

Chanos fishponds utilize tidal fluctuations for letting in and draining out water in the ponds when needed in management. Under a given tidal characteristic such as in the Philippines, determination of the appropriate elevations for milkfish pond projects based on the tidal datum can be made. The method of determination and its application are discussed.

Ronquillo, I.A., E. Villamater, *Observations on artificial feeding of bangos fry, Chanos chanos (Forskal).* Philipp. J. Fish. 5(2):103-112. 1957.

The greatest mortality in bangos fry under cultivation occurs during the first few weeks of life in the ponds. It has been found that the greatest single factor necessary for survival during the delicate early stage of their life is the availability of readily assimilable protein of high nutritive value.

Ronquillo, I.A., E. Villamater, and H. Angeles, *Observations on the use of terramycin and vigofac enriched diet on 'Bangos' fry, Chanas chanos.* Proc. Indo-Pacific Fish Coun. 7(II), 1957.

A preliminary study on the use of antibiotics and animal protein factors in the feed for 'bangos' fry was made under controlled conditions.

Ronquillo, I.A. and deJesus, A. *Notes on growing of lab-lab in bangos nursery ponds.* Phillipp. J. Fish. 5(2):99-102. 1959.

Difficulty in growing lab-lab used for feeding milkfish fry and fingerlings has been solved by the use of organic fertilizers.

Sulit, J. and others, *Fertilization of bangos nursery ponds with commercial chemical fertilizer.* Philipp. J. Fish. 5(2):125-133. 1959.

This is a study in ways to improve and increase production and reduce mortality of fry in the nursery pond. Experiments were conducted by using complete chemical inorganic fertilizer to replace the usual practice of depending on the natural growth of lab-lab produced by the fermentation of unused refined rice bran used as feed of the fingerling.

Tang, Y.A. and S.H. Chen, *A Survey of the algal pasture soils of milkfish ponds in Taiwan.* FAO Fish. Rep. 44(3):198-209. 1967.

The total area of milkfish ponds in Taiwan at present is about 18,000 ha.

Tang, Y.A. and T.L. Hwang, *Evaluation of the relative suitability of various groups of algae as food of milkfish in brackish water ponds.* FAO Fish. Rep. 44(3):365-372. 1967.

An evaluation of the relative suitability of the four major groups of brackish water pond algae as milkfish food was made by digestion experiments, observations on their suitability as food and effect on proper pond managment.

6. EELS

Boetius, I. and J. Boetius, *Studies in the European eel, Anguilla anguilla (L). Experimental induction of the male sexual cycle, its relation to temperature and other factors.* Medd. Danmarks Fisk. Havunders. 4(8/11):339-405. 1967.

Male silver eels were injected with 250 IU of HCG (human chorionic gonadotropin) weekly in sea water, 14°C. A complete sexual cycle was induced and characterized by seven developmental stages.

Eales, J.G. *A bibliography of the eels of the genus Anguilla.* Fish. Res. Board. Can. Tech. Rep. 28. 1967. p. 171 (1).

Hasegawa, Hitoshi, *On the culture of imported European glass-eels.* Bull. Shizuoka Prefectural Fish. Expt. Station #5. 1972.

Koops, H. *Ein Futterungsversuch mit Satzaalen in der Flussteichwirtschaft Muden an der Mosel [A feeding experiment with small eels in the river-pond economy of Muenden on the Mosel.]* Arch. Fischereiwiss. 17(1):36-44. 1966. English summary.

An experiment was made to feed small eels in a pond.

_____. *Die Aalproduktion in Japan. [The production of eels in Japan].* Arch. Fischereiwiss. 17(1):44-50. English summary. 1966.

A survey of the Japanese eel-fishery and eel-pond culture, as published in Japanese fishery statistics is given. The main eel-fisheries are in middle and southern Japan. The eels are fed with fish and residual fish after processing. The food quotient is about 8-9.

_____. *Feeding of eels in ponds.* FAO Fish. Rep. 3(44):359-364. Oct. 1967.

It is possible to feed elvers and eel fingerlings in ponds with minced fish and shrimps or fresh meat. The food quotient for elvers is lower than 10 and for fingerlings also it may be about the same. The results of experiments described in this report show that commercial pond feeding of eels is feasible.

Lauterbach, R. *An attempt to fatten eels in net enclosures.* Fisch. Forsch. 6(1):87-91. 1968. 14 ref.

In a feeding experiment in net enclosures the average weight of *Anguilla anguilla* was raised from 71 to 155 g. The main additional growth took place over three months during which time the average weight almost doubled. The method is discussed.

Meske, C. *Rearing of elvers in aquaria.* Arch. Fischereiwiss. 20:26-32. 1969.

Eels were raised in aerated warm-water tanks and fed with either fresh-fish/dry-food mix or on a dry-food/added crude protein mixture. All eels, particularly those fed on dried food only, showed good growth rates, with growth capacity vastly increased over that observed under natural conditions. The considerable size deviation among eels of the same age group and the feasibility of practical eel farming under warm-water conditions are discussed.

7. YELLOWTAIL

Yamamoto, T. *L'aquaculture: L'elevage des serioles au Japon.* (in French) France Peche (166):106-109, 1972.

Seriola aquaculture in Japan is described. There were three types of fish ponds used: embanking ponds, ponds closed by nets, and floating net ponds. The floating net ponds were found to be the most efficient type. They were kept six months before selling. The growth rate and breeding costs are given.

8. SHRIMP

Allen, D.M. *Shrimp farming.* U.S. Bur. Comm. Fish. Fish. Leafl. 551. 29 ref. November 1966. 8 p.

In the United States, natural populations of shrimp occur in estuaries along the Gulf and South Atlantic coasts. This region appears well suited for shrimp farming, but commercial ventures in this field have been unsuccessful. If shrimp culture problems relating to seed supply, growth, survival, and harvesting can be resolved, and proven methods followed closely, Gulf and South Atlantic coastal marshes may support a new industry.

Anderson, W.W. *Contributions to the life histories of several penaeid shrimps (Penaeidae) along the South Atlantic coast of the United States.* U.S. Fish Wildlife Serv. Spec. Sci. Rep. Fisheries No. 605. 24 p. 1970.

Shrimp, the most valuable fishery resource of the South Atlantic coast of the United States, contributed about 40 percent of the $27 million exvessel value of all fishery landings in the area in 1966. Three species of shallow-water penaeid shrimps are of greatest commercial importance: white shrimp, *Penaeus setiferus;* brown shrimp, *P. aztecus;* and pink shrimp, *P. duorarum.* The shrimp fishery is reviewed for trends in yield for the area as a unit, by States, and by species, for the 10-year period 1958-67. A trend toward steady decline in total shrimp landings is indicated. During studies on the white shrimp along the South Atlantic coast of the United States in 1931-35, data were obtained on the brown shrimp; the sea bob, *Xiphopeneus kroyeri;* and *Trachypeneus constrictus.* Observations were also made on the pink shrimp from operations of the Bureau of Commercial Fisheries R/V Oregon off northeast Florida near Cape Kennedy in 1965-67. This report presents size distribution, ovary development, and sex ratios of the several species of shrimp, and includes limited information on spawning season.

Anonymous. *Kaap-Kunene plan prawn farm on Kosi Bay Lakes.* The South African Shipping News and Fishing Industry Review, 26(5):64-65, 67. 1971.

This project appears to form the basis for a whole new industry which will do little to spoil the area, which may enrich the marine environment, and which will offer badly needed employment to several hundreds of the local inhabitants. The basis of the method is to increase the zooplankton, phytoplankton and algae content of the water to which prawns are introduced. This is achieved by adding organic fertilizer to the water. Under natural conditions, with the addition of organic fertilizer, an average production of 200 g per m^2 can be achieved. Should the Ku-Mpungwini project prove a success, it could lead not only to prawn cultivation in another area, but also to the cultivation of other types of fish.

_____. *Western Atlantic shrimps of the genus Penaeus.* Fish. Bull. U.S. Fish and Wildlife Serv. 67(3):577-590. Bibliogr. 1969.

An extensive bibliography from 1790 to 1968 on *Penaeus* including taxonomy, ecology and life histories.

_____. *Aquaculture: the new shrimp crop.* Sea Grant Inf. Leafl. 1970. Miami University. Sea Grant Institutional Program.

The author outlines the ecological and economic feasibility of marine aquaculture. He reviews the progress that has been made in culture of shrimp in Southeast Asia, Japan and the Philippines. In the United States, several organizations have undertaken programs of shrimp culture. These organizations are listed and their programs are described.

_____. *Prawns in culture.* Nature (London) 223(5210):999-1000. 1969.

The Ministry of Agriculture's Fisheries Experiment Station in Conway, North Wales, reports successful culture of adult prawns (*Palaemon seratus*) from eggs. Previously, breeding was dependent on wild stock and was subject to adverse effects of extremely cold winters. The environmental conditions, feeding and behavior problems, and subsequent size of harvest are discussed. It has been demonstrated that it is possible to obtain 90 percent survival if larvae through metamorphoses to adulthood.

Broom, J.G. *Pond culture of shrimp on Grand Terre Island, Louisiana, 1962-1968.* Proc. Gulf Caribbean Fish. Inst. 21:137-151. 1969.

Pond culture experiments were started in the spring of 1962 at Grand Terre Island, Louisiana. Brown shrimp, *Penaeus aztecus*, and white shrimp, *Penaeus setiferus*, were cultured in 0.25 acre ponds. Construction details of the ponds used in this study are discussed. Juvenile and postlarval shrimp, obtained from several sources, were stocked at different rates and several types of feeds were used. The shrimp were fed a specific percentage of their body weight daily. A sample of 50 shrimp was taken each week, weighed and returned to the ponds. Feeding rates were adjusted to the increased growth without regard to possible mortalities. The best feeding rates were 5 percent and 10 percent at a stocking rate of 20 thousand shrimp per acre. Implications of feeding, including maximum possible feeding rates and amounts, are discussed. Production ranged from 40 to 809 pounds per acre and feed conversion ratios from 1.7 to 9.7. Salinities fluctuated between 16 and 35 ppt while temperatures varied between 8° and 37°C during the study periods.

Caces-Borja, P. and S.B. Rasalan. *A Review of the culture of sugpo, Penaeus monodon Fabricius, in the Philippines.* FAO Fish. Rep. 57:111-123. 1968.

Penaeus monodon was cultured in association with the culturing of *Chanos chanos* Forskat (bangos) in the Philippines. A sugpo fry fishery has not been developed since the fry are easily recognizable by the dark brown pigment. Sugpo fry migrate to the shore and into brackish waters for feeding and shelter. The methods of collecting (fish lures, dip net), selection of site, construction, and preparation of a pond, and the raising and harvesting of sugpo are described. The problems of cultivation are discussed, the main one being the lack of fry for stocking. Other problems are low rate of survival and the difficulty of cropping. Comparison of the production of a pure bangos, a mixed bangos and sugpo, and a pure sugpo culture showed that the latter gave highest production.

Campbell, K.C. *Prawn farming in the Far East.* World Fish. 18(1):41-43. 1969.

Prawn farming in Japan, Korea, Malaysia, Hawaii and the Philippines is described in terms of the species used and the methods of rearing from the egg up to market size. The problem of predators and suitable food is also considered. The techniques used are relatively sophisticated, particularly when compared to those employed in American lagoons.

Chin, Edward and D.M. Allen. *A list of references on the biology of shrimp (family Penaeidae).* U.S. Fish and Wildlife Service Spec. Sci. Rept. 276, 1959.

Choudhury, P.C. *Complete larval development of the Palaemonid shrimp Macrobrachium acanthurus (Wiegmann, 1836), reared in the laboratory.* Crustaceana 18(2):113-132. 1970.

Extensive descriptions of all the larval stages of *M. Acanthurus* with figures aiding identification of stages.

Choudhury, P.C. *Complete larval development of the Palaemonid shrimp Macrobrachium carcinus (L.), Reared in the laboratory (Secapoda, Palaemonidoe)* Crustaceana 20(1):51-69, 1971.

Describes the complete larval stages, with extensive figures aiding keying out to stage.

Cook, H.L. and M.A. Murphy. *The culture of larval penaeid shrimp.* Trans. Amer. Fish. Soc. 98:751-754. 1969.

Large numbers of penaeid shrimp were reared to postlarvae from eggs spawned in the laboratory. Rearing containers were four 1040-liter tanks and one 1890-liter tank.

Cook, H.L. and M.A. Murphy. *Rearing penaeid shrimp from eggs to postlarvae.* Contrib. U.S. Bur. Comm. Fish. Biol. Lab. 209. 1965.

A description is given of the physical facilities in which mass cultures of penaeids have been reared from eggs to postlarvae. The metal chelator EDTA was added to the water in which the shrimp were grown. Larvae of *Penaeus aztecus* developed more rapidly at 30°C than at lower temperatures. Salinity varied from 20.5 ppt to 36.0 ppt during rearing trials in which *P. aztecus* larvae were reared to postlarvae. Addition of mixed algal cultures as food gave better survival than additions of their individual components. EDTA was used as an additive to filtered sea water to grow a diatom, *Skeletonema sp*, in mass culture, as food for larval shrimp.

Costello, T.J. *Pink shrimp life history.* Circ. U.S. Fish and Wildlife Serv. 161:35-37. 1963.

Projects were designed to provide information on the life history of the pink shrimp (*Penaeus duorarum*), which is commercially important, by the Bureau of Commercial Fisheries in south Florida. Considerable data on the identification and distribution of the pink shrimp have been collected as well as on migration, growth, mortality rates, and geographical range of juvenile and adult stocks.

Cummings, W.C. *Maturation and spawning of the pink shrimp, Penaeus duorarum Burkenroad.* Trans. Am. Fish. Soc. 90(4):462-468. 1961.

A description is given of the anatomy of the reproductive system of male and female pink shrimp, *Penaeus duorarum*. Four stages of female maturation (undeveloped, developing, nearly ripe, and ripe) were observed using ovum size frequency, gross observation, and ratio of gonad weight to tail weight. Size of female at first sexual maturity is estimated to be about 22 millimeters (carapace length). Months of higher spawning activity are April through July. Throughout the study five ripe females were observed; four of these were found during the months of highest spawning activity. It is probable that the female pink shrimp spawns more than once during a lifetime. There is positive correlation between size of shrimp and number of ova spawned. Although temperature is the only environmental factor measured, there is reason to believe that spawning is closely associated with the annual rise and fall of bottom temperatures.

Dall, W. *Food and feeding of some Australian penaeid shrimp.* FAO Fish. Rep. 2(57):251-258. Oct. 1968.

Food of the Australian commercial penaeid shrimps, *Penaeus esculentus, P. merguiensis, P. plebejus, Metapenaeus bennettae,* and *M. macleayi,* was found to consist of the remains of small animals and a large amount of unrecognizable material. It is suggested that the latter forms the main component of the diet, and that shrimps derive this by browsing on the microorganisms (bacteria, algae, and micro-fauna) which grow on the surface of the substrate. Gross structure and function of the gut are described. Digestion and assimilation were found to be largely completed within six hours at 20°C, and this is attributed to the finely divided nature of the food. The need for detailed study of the productivity of the macroorganisms of the estuarine and inshore substrate is discussed.

De La Bretonne, L.W. and J.W. Avault. *Shrimp mariculture methods.* Amer. Fish. Farmer 1:8-11. 27. Nov. 1970.

Natural estuarine areas, such as the Louisiana State University's research facility, have been the prime location for mariculture studies of shrimp. Shrimp culture in artificial ponds will depend on alteration of naturally occurring environmental factors in order to determine optimum conditions for rearing shrimp. Future research must be directed toward the study of cover grasses, fertilization, and effects of water exchange in ponds. Information and data must be obtained on shrimp behavior and migratory patterns, optimum temperature and salinity, stocking rates, pond construction, commercial feeds, and disease problems. Proper development and operation of the pond culture of shrimp can serve to stabilize the commercial shrimp market so that a predictable shrimp crop can be harvested each year.

Dobkin, S. *Abbreviated larval development in caridean shrimps and its significance in the artificial culture of these animals.* FAO Fish. Rep. 57:935-946. 1969.

The author's studies on the larvel development of one penaeid and more than 40 caridean shrimps have produced information applicable to the development of successful culture techniques.

Ewald, J.J. *The laboratory rearing of pink shrimp Penaeus duorarum Burkenroad.* Bull. Mar. Sci. 15(2):436-449. 1965.

In August and September 1963, the first successful attempts were made to rear pink shrimp from eggs spawned in the laboratory through all larval and post-larval stages. Ripe female shrimp were collected from fry-net trawls on the Dry Tortugas fishing grounds and placed in individual aquaria. Eggs were collected and transferred to 8-inch diameter finger bowls and development was allowed to proceed until the first protozoeal stage. Individual protozoeae were placed in compartments of plastic fishing tackle boxes containing fresh sea water. Water and food were changed each day. A mixture of unicellular algae and yeast was used as food. Some larvae reared in Gulf Stream (Florida Current) water reached adulthood whereas none of those reared in Biscayne Bay water survived. Optimum temperature for rearing was 26°C. General morphological development followed closely that given by Dobkin (1961); however, number of intermolts and time of development varied considerably with individuals. Fastest development from eggs to postlarvae occurred in 15 days.

Forster, J.R.M. *Further studies on the culture of the prawn, Pelaemon serratus Pennaut, with emphasis on the postlarval stages.* Fish. Invest. 26(6):1-40. 1970.

Following the work of Reeve (1969) the conditions for optimum growth and survival of juvenile prawns was studied.

Fujinaga, G. *Facilities for shrimp farming* (in Japanese). Tokyo Koho (patent Agency Bulletin), 20 July, Patent Application No. 33-28986:1-2. 1960.

Translated by JRAS for BCF, October 1970, pp. 5, typescript. Available on loan—BCF. Washington, D.C.

Fujinaga, G. *Techniques of shrimp farming during zoea, mysis and postlarval stages* (in Japanese). Tokyo Koho (Patent Agency Bulletin) 11 May, Patent Application No. 34-27796:1-3. 1961.

Translated by JPRS for BCF, October 1970, 10 p., typescript. Available on loan—BCF, Washington, D.C.

Fujinaga, M. *Culture of Juruma-shrimp (Penaeus japonicus).* Curr. Aff. Bull. Indo-Pacific Fish. Coun. (36):10-11. 1963.

Spawning and culture of larvae in aerated aquaria (25°C) are described. An adult of 25 g produced about 1,200,000 eggs which hatched in 14 hours. Naupilius stages require 36 hours during which six molts occur. Zoea stages were fed *Skeletonema* for four days at which time the mysis stage was reached. Mysis larvae were fed brine shrimp during the three-day stage and during further postlarval development. Young prawns are attained after 15-20 days as postlarvae. Prawns can be fed on any animal material, bivalves being favored. About 20,000 young prawns are produced from a single spawner. They reach commercial size in a year in outdoor ponds to which they are transferred after four or five days in the post-larval stage.

Fujinaga, M. *Kuruma shrimp (Penaeus japonicus) cultivation in Japan.* FAO Fish. Rep. 57:811-832. 1969.

The history of Karuma shrimp cultivation in Japan is traced from the first attempts to rear *Penaeus japonicus* in the laboratory in 1933 to the large-scale commercial rearing practiced in 1967. Successful rearing on a large scale was first achieved in indoor tanks, using cultures of the diatom *Skeletonema costatum* as food for the zoeal stages, *Artemia* nauplii for the mysis and post-larval stages and crushed clam meat for the juvenile stages. The scale of operations has been greatly increased and the cost of production lowered by using large outdoor tanks filled with natural sea water to which nutrient salts are added. The resulting bloom of phytoplankton makes the separate culture of diatoms unnecessary, and wild zooplankton and benthos develop to a considerable degree, so that only relatively small amounts of *Artemia* and clam meat need be added. Current research is largely directed to finding suitable foods other than potential human foods, for juvenile shrimp.

George, M.J. *On the breeding of penaeids and the recruitment of their postlarvae into the backwaters of Cochin.* Indian J. Fish. 9(1):110-116. 1962.

A quantitative study of the postlarvae of penaeid prawns *M. dobsoni, M. Monoceros,* and *P. indicus* in the backwater plankton of Ernakulam is made for five years from 1956 to 1960. The size ranges of the postlarvae are described. Possible relationship of seasonal fluctuations in recruitment of these and salinity of the water is also indicated.

George, M.J., K.H. Mohamed, and N.N. Pillai. *Observations on the paddy-field prawn filtration of Kerala, India.* FAO Fish. Rep. 2(57):427-442. October 1968.

Experiments were conducted to determine whether culture methods could be advantageously introduced into the existing prawn filtration practices in the paddy fields of Kerala, on the southwest coast of India. The pattern of the fishery is changing, with the demand for small prawns decreasing and that for large prawns increasing. The experiments indicated that culturing of juvenile prawns for about a month resulted in relatively better catches of large-sized prawns that could be obtained by cultivation for longer periods. The yield of prawns appeared to be better during the spring tide period associated with the full moon than that associated with the new moon; this is correlated with the increased tidal gradient at full moon. The fishery mainly concerns four species of penaeid prawns, namely, *Metapenseus dibsoni, M. monoceros, M. affinis,* and *Penaeus indicus.* Recruitment to the fishery is continuous with a peak in March or April. The majority of the juvenile prawns brought into the field by a particular tide do not seem to move out of it when the tide recedes.

Holthuis, L.B. and H. Rosa. *List of species of shrimps and prawns of economic value.* FAO Fish. Tech. Pap. 52. 1965. 21 p.

Lists species in systematic order with notes on their geographic distribution and commercial value. Common names of some species are given.

Hudinaga, M. and Z. Kittaka. *Studies on food and growth of larval stage of a prawn, Penaeus japonicus, with reference to the application to practical mass culture.* (in Japanese). Inform. Bull. Planktol. Japan 13:83-94. 1966.

Translated by JPRS for BCF, July 1970, 29 p., typescript. Available on loan—BCF, Washington, D.C.

Huner, J.V. *Use of Fintrol-5 to control undesirable fishes in shrimp-oyster ponds.* Proc. La. Acad. Sci. 81:58-61. 1968.

At a concentration of 5 ppm (5 ppb antimycin A, the active ingredient) Fintrol-5 virtually eliminated *Cyprinodon variegatus* from shrimp-oyster ponds. It had no apparent effect on shrimp, oysters, and other invertebrates in the ponds.

Idyll, C.P., D.C. Tabb, and W.T. Yang. *Experimental shrimp culture in southeast Florida.* Proc. Gulf Caribbean Fish. Inst. 21:136. 1969 (abst. only)

Long-term research has been commenced on intensive culture of pink shrimp, *Penaeus duorarum*, under the auspices of the Armour and United Fruit Companies and the National Science Foundation "Sea Grant" program. Facilities for this research have been constructed on the grounds of an electrical generating plant owned by Florida Power and Light Company in south Dade County, Florida. The present 5-acre site contains four quarter-acre, two half-acre and one 1-acre ponds. In addition, construction is nearing completion on 16 outdoor concrete larval rearing tanks each having a capacity of 20 metric tons of sea water. A laboratory has been built where research is being carried out on larval culture and propagation of larval food organisms. Initial experimentation has involved the use of wild pink shrimp as a source of eggs, but a selective breeding program is planned. Shrimp have been successfully reared from the egg to pond stocking size during the first year of the program.

Joyce, E.A. and B. Eldred. *The Florida shrimping industry.* Educ. Ser. Fla. Board Conserv. 15:1-47. 1966.

Major commercial species, commercial shrimping, minor shrimp species, bait shrimping, shrimp farming, and future of the industry are the topics covered. The author outlines the major problems confronting the shrimp farming industry, the major one being the securing of sufficient numbers of larvae for stocking purposes. Impoundment of wild postlarvae seems to be the most feasible from the author's view at this time.

Karim, M. and D.V. Aldrich. *Influence of diet on the feeding behavior, growth and thermal resistance of postlarval Penaeus aztecus and P. setiferus.* Texas A&M Univ. Sea Grant Program. September 1970. 80 p.

Laboratory studies were conducted on the effects of diet on the food preference, survival, growth and temperature tolerance. *Artemia* and five artificially compounded foods were tested. Results indicated that initial diet preference, survival, growth and resistance to high temperature are independent qualities in terms of these two species of shrimps.

Kesteven, G.L. and T.J. Job. *Shrimp culture in Asia and the Far East. A preliminary review.* Proc. Gulf Carib. Fish. Inst. (1957) 10:49-68. 1958.

This paper deals with cultivation of shrimp in Asia. Some of the examples are not of true cultivation, but are industrial practices which are included because of their relations with true cultural practices and because they furnish information on shrimp in captivity. Two problems must be considered in culturing an organism: (1) the biology and life history of the animal, and (2) ecological, economic and dietary requirements of the consumer. Species being cultivated are listed with the common names and scientific names, along with the countries where cultured. Habitats and bionomics of important species are given. Descriptions of shrimp farms with their size, layout, ecology and equipment are included. Also described are farming techniques, fry collection, rearing fry, nursery upkeep, transplantation, care of stock and harvesting.

Kim, K.D. *Studies on the artificial culture of Penaeus orientalis Kishinouye.* FAO Indo-Pac. Fish. Council, Tech Paper 19.

Survival from hatching to post-larvae is only 6 percent, and survival from this stage on is 30 percent. Experiments in artificial hatching are described.

Kirkegaard, I. and R.H. Walker. *Synopsis of biological data on the tiger prawn, Penaeus esculentus Haswell, 1879.* Aust. Commonw. Sci. Ind. Res. Organi. Div. Fish Oceanogr. Fish. Synop. 3. 1969. 30 p.

Biological data on the tiger prawn are presented in terms of the organism's nomenclature, taxonomy, morphology, distribution, bionomics and life history, population, exploitation and protection and management. The section on nomenclature, taxonomy, and morphology is the most extensive, including such areas as cytomorphology, protein specificity and subspecies.

Kow, T.A. *Prawn culture in Singapore.* FAO Fish. Rep. 2(57):85-93. Oct. 1968.

The method of prawn culture in Singapore differs from the one used in the Philippines mainly in that young prawns are not stocked in the pond by operators but are brought in by tidal flow. The method appears simple, yet many operators have failed to achieve success because they are not sufficiently experienced in the selection of the site, and in the construction, operation and maintenance of the pond. In this paper these methods, which are largely empirical, are described and an attempt made to explain the basic principles involved.

Kutkuhn, J.H. *Dynamics of a penaeid shrimp population and management implications.* Fish. Bull. U.S. Fish and Wildlife Serv. 65(2):313-338. 1966.

In assessing present utilization of a stock of pink shrimp (*Penaeidae*) that supports an important commercial fishery in the eastern Gulf of Mexico, the interaction of population growth and mortality is critically analyzed. Estimates of the parameters involved were secured through a mark-recapture experiment wherein a biological stain served as the marking agent. The experiment was oriented in space and time so that exploitation of the marked population, which initially consisted of individuals uniform in size, provided measures of growth and mortality in the parent age group during and immediately following its transition from prerecruit to postrecruit status. Throughout the experiment, the entire stock as well as the marked population were heavily fished. Upon examining the question of whether or not the fishery's production could be improved by postponing the start of fishing until the shrimp reach a size greater than the 70 headless-count designation now generally viewed as a practicable minimum, it was noted that expected growth, although relatively high, would be insufficient to offset substantial losses due to expected natural mortality. Even with a moderate increase in growth rate, an appreciably reduced natural mortality would have to be indicated before such a move could be considered feasible. Maximum potential yield in both weight and value is obtained with the minimum acceptable size that the fishery currently imposes.

Kutty, M.N. *Oxygen consumption of the prawns Penaeus indious H. Milne Edwards and Penaeus semisulcatus de Haan.* FAO Fish. Rep. 3(57):957-969. 1969.

The routine oxygen consumption of *Penaeus indicus* and *Penaeus semisulcatus* (acclimation and test in sea water at 30°C), starved for 5-10 days, declined sharply by about the second day of starvation, and no marked change from the reduced rate was observed during the subsequent days of starvation in either species. The reduced level of metabolism attained due to starvation was 0.404 mg/g/h, being 32 percent less than the metabolic rate for the first day of starvation in the case of *P. indicus*, whereas the corresponding values for *P. semisulcatus* were 0.151 mg/g/h and 57 percent. When the amount of available dissolved oxygen is the same, *P. indicus* and *P. semisulcatus* starved for two days or over can be expected to survive for 1.7 and 2.3 times as long, respectively, as those not starved, under the conditions of the present tests. The present regression of log oxygen consumption on log weight in *P. indicus* has a slope of 0.501. While the highest routine metabolic rate of *P. indicus* declined with the decrease in ambient oxygen concentration, such oxygen dependence need not be exhibited in all cases. There is possibly an increase in the standard metabolic rate of *P. indicus* at an oxygen level of 1-2 ppm when the prawns are allowed to bring down the ambient oxygen concentration by their own respiration.

Lindner, M.J. and H.L. Cook. *Progress of shrimp mariculture in the United States.* FAO Fish. Rep. 71.1 153 p. 1969.

Larvae of the three commercially important penaeid shrimp native to the U.S.—*Penaeus aztecus, P. setiferus, P. duorarum*—have been cultured in the laboratory. Within the next year at least 12 government or university agencies will be conducting research; at least four private concerns are now in operation or planning to start in the near future, and four industrial groups are supporting research.

Marifarms, Inc. *Artificial shrimp culture.* U.S. Patent 3,473,509 (21 Oct. 1969).

Shrimps are cultured by depositing pregnant females of the species in a darkened tank containing aerated water, and allowing one or more of them to spawn before removing all the adults and finally feeding the larvae over a period of weeks with phytoplankton, zooplankton, brine-shrimp eggs, and crushed mussels, clams and bivalve meat. The temperature should be kept to about 25°C throughout, except in the final stages when the shrimps (20 mm in length) are transferred to an outside pond. The shrimps are marketed when they reach a weight in excess of 20 grams. The object, throughout, is to supply food to the larvae at a fast enough rate to prevent them eating one another, while maintaining them in a healthy state and protecting them from outside prey. A female shrimp can spawn as many as a million eggs, and it is claimed that 50-80 percent of these can be successfully reared.

Panikar, N.K. *Osmotic behavior of shrimps and prawns in relation to their biology and culture.* FAO Fish. Rep 2(57):527-538. Oct. 1968.

The prawns which are useful for cultural purposes belong mostly to the decapod families *Penaeidae* and *Palaemonidae*. Most penaeids are marine prawns which migrate to estuaries and brackish water in their young stages but go back to the sea to breed. A small number of them breed in coastal inlets and others are exclusively marine. Species of marine *Palaemon* are highly adaptable to lower salinities. The habitat of *Palaemonetes* ranges from sea water to freshwater while *Macrobrachium* is largely a freshwater genus. Species of *Macrobrachium* include those which migrate from freshwater to brackish water during the breeding season. The marine penaeids and palaemonids show capacities for hypo-osmotic regulation in brackish and freshwater. They have not developed extreme specialization as freshwater inahbitants in that developed extreme specialization as freshwater inhabitants in that established freshwater crustacea and they do not produce hypotonic urine for conserving salts. These features have endowed these prawns with unusual adaptational abilities to live in variable surroundings, although each species has its own optimal range. There is close correlation between euryhalinity and hyper-osmotic adaptation. There is a relationship between osmotic behavior and temperature, and a combined influence of temperature and salinity is also in evidence. A fuller knowledge of isosmotic levels and critical evaluation of the influence of environmental conditions on osmotic and ionic behavior would help to rationalize prawn and shrimp cultural practices. The paper discussed the distribution of prawns and shrimps, their value in culture, existing knowledge of their osmotic properties, influence of salinity and temperature on their physiology and related problems.

Rajyalakshmi, T. *Observation on the embryonic and larval development of some estuarine palaemonid prawns.* Proc. Natl. Inst. Sci. India (B) 26:395-408. 1960.

An account of the embryonic development, hatching and the structure of the early larvae of *P. malcomsonii, P. Rudis, P. scabriculus,* and *P. mirabilis* is given. A comparison between the different species described, shows certain differences in size, time taken for development and hatching and the time of appearance or functioning of various embryonic structures. Differences in the nature of chromatophores and their arrangement in the first stage larvae of all the species studied, have been described and the salient points of variation in the morphometric characters that distinguish the larvae of different species are discussed. The second stage larva of *P. mirabilis* is described in detail.

Rajyalakshmi, T. *Studies on maturation and breeding in some estuarine palaemonid prawns.* Proc. Natl. Inst. Sci. India (B) 27(4):179-188. 1961.

An account of the studies on ova-maturation, fecundity and spawning characteristics of two commercially important species of *Palaemon,* namely, *P. carcinus* and *P. mirabilis,* is given. In *P. carcinus* breeding period is confined to 6-7 months, from December to July, with a peak period from March to May. An individual appears to breed more than once in a season. The species seems to congregate at the middle zone of the Hooghly estuary for maturation of ovary, spawning and hatching. *P. mirabilis* is an almost perennial breeder. In this species, though its maturation, spawning and early development of its eggs appear to occur all along the river, the later development and hatching seem to occur only at the lower reaches of the middle zone and lower zone.

Reeve, M.R. *The laboratory culture of the prawn Palaemon serratus.* Fish. Invest. Lond. (Ser. 2) 26(1):1-38. 1969.

The laboratory culture of *P. serratus* is described and discussed. Subjects covered include: the maintenance of laboratory stock with respect to the quality of breeding stock; molting; mortality in stock tanks; length and wet and dry weights; oxygen consumption; and ammonia production. The growth and survival of young stages under different rearing conditions (temperature, darkness, water quality, agitation and salinity) is considered and the effect of density on cannibalism is discussed. The results showed that the optimum temperature range for growth was 20-25°C. The optimum salinity was that of normal sea water. Density and survival were inversely related. *Artemia* nauplii proved to be the only adequate food for the prawn larvae and differences in the food value of *Artemia* stocks were observed. Mussel meat is considered as a food for older animals. Long-term rearing of *P. serratus* and *P. elegans* indicated that growth could be doubled compared with that in the natural environment.

Reeve, M.R. *The suitability of the English prawn, Palaemon serratus (Pennant) for cultivation—a preliminary assessment.* FAO Fish. Rep. 3(57):1067-1073. 1969.

The paper reports the results of a preliminary study into the culture of *P. serratus*. All parts of the life cycle could be completed in the laboratory, including maturation of the gonads, copulation, spawning, egg carriage, hatching and growth and metamorphosis of larvae, and growth of postlarvae and juveniles. The cultivation of *P. serratus* certainly appears to be biologically feasible but the economic feasibility may be much less certain.

Sick, L.V., J.W. Andrews, and D.B. White. *Preliminary studies of selected environmental and nutritional requirements for the culture of Penaeid shrimp*. Fish. Bull. U.S. Dept. of Commerce 70(1):101-110. 1972.

The types of substrate, aeration and stocking density for high-density culture of Penaeids are discussed. Diets of semi-purified pellets reported on and the nutritional requirements are described.

Subrahmanyam, C.B. and C. H. Oppenheimer. *Food preference and growth of grooved penaeid shrimp*. Proc. Food-Drugs. Sea 1969:65-76. Washington, D.C. Mar. Technol. Soc., 1970.

The authors found, in some preliminary experiments, that penaeid shrimp can be raised on pelletized food made of fishmeal and other ingredients. The present investigation has been designed to find out the relative efficiency of three types of pelletized food in the growth of grooved shrimps. The success of such experiments may have far-reaching implications, especially in the present context of increasing consciousness of maricultural practices.

Tabb, D.C., C.P. Idyll, W.T. Yang, and E.S. Iversen. *Progress on shrimp and pompano culture*. Abstr. Amer. Fish. Soc. Annu. Meet. 98:17-18. 1968.

The Institute of Marine Sciences, University of Miami, has embarked on long-term research into intensive culture of pink shrimp, *Penaeus duorarum*, and pompano, *Trachinotus carolinus*, under the auspices of the National Science Foundation "Sea Grant" program. A selective breeding program will go hand-in-hand with the rearing and spawning of their own adult brook stock research.

Tournier, H., C. Juge, and C. Carries. *Conditions for the acclimitization of the shrimp Penaeus kerathurus and P. japonicus in the waters of the Languedoc coast*. (in French). Science et Peche No. 213:1-13. 1972.

At the Sete laboratory of the Institut scientifique et technique des peches maritimes has initiated in 1965 a study of the conditions of life in captivity of *Penaeus kerathurus* and *P. japonicus*.

Villella, J.B., E.S. Iversen, and C.J. Sindermann. *Comparison of the parasites of pond-reared and wild pink shrimp (Penaeus duorarum Burkenroad) in South Florida*. Trans. Amer. Fish. Soc. 1970.

Parasites, with percentage incidence, given for pond-reared (at Turkey Point, Florida) and wild (Biscayne Bay, Florida) are given. Absence and reduced incidence of certain parasites in pond-reared *P. duorarum* is discussed.

Wheeler, R.S. *Experimental rearing of postlarval brown shrimp to marketable size in ponds.* Comm. Fish. Rev. 29(3):49-52. Mar. 1967. Also issued as U.S. Fish and Wildlife Service Separate 785.

Two rearing methods were used. In one pond, shrimp were fed a prepared diet, and filtered sea water was pumped through the pond at a rate of 60 gallons per minute. During a 95-day period, shrimp showed continuous growth and attained an average length of 97.4 mm (about 106 tails per pound); the projected production was 234 pounds per acre. In the second pond, commercial fertilizer was applied to stimulate plankton growth, and the water was maintained in a static condition. During a 4-month period, shrimp attained an average length of 80.0 mm (about 200 tails per pound) and had a projected production of 45 pounds per acre. Coefficients of condition showed that shrimp held in the circulating-water pond maintained, in general, a good state of relative well-being; those held in the static-water pond could not.

Wheeler, R.S. *Culture of penaeid shrimp in brackish-water ponds, 1966-67.* Proc. S.E. Assoc. Game and Fish Comm. 22:387-391. 1968.

Young shrimp have shown rapid growth in brackish-water ponds which had been fertilized, but to which no supplemental feed was added. In 1966, white shrimp (*Penaeus setiferus*) were stocked at the rate of nine shrimp per square meter of bottom in a pond that had been fertilized with chicken manure. In 1967, brown shrimp (*P. aztecus*) were stocked at a rate of 22 shrimp per square meter of bottom in one pond that was fertilized with rice husks and in another that was not fertilized. In both experiments initial growth was rapid; the shrimp attained bait size (75 to 93 mm total length) in five to seven weeks. This rapid growth was followed by a period of slow growth. In 1967, supplementary feeding produced additional gains after growth had nearly ceased. Survival of the white shrimp was 84 percent, whereas survival of the brown shrimp was 23 percent in the untreated pond and 31 percent in the fertilized pond. Oxygen deficiencies caused by dense blooms of phytoplankton during the 1967 experiment resulted in several mass mortalities.

Zein-Eldin, Z.P. *Shrimp physiology.* Circ. U.S. Fish and Wildlife Serv. 129:44-48. 1961.

Specimens of both white and brown shrimp have been tested for oxygen requirements, mineral and vitamin needs, response to artificial and natural diets, and growth and survival in artificial and natural sea water. Tables showing the composition of the artificial diet and the artificial water are presented.

Zein-Eldin, Z.P. *Effect of salinity on growth of postlarval penaeid shrimp.* Biol. Bull. 125(1):188-196. Aug. 1963.

The effect of salinity on the growth and survival of postlarvae of white, *Penaeus setiferus*, and grooved shrimp, *P. Aztecus* or *P. duorarum*, has been studied in the laboratory. Growth rate did not differ significantly among shrimp held at 2, 5, 10, 25, or 40 0/00. Survival was generally excellent at all salinity levels tested, including 40 0/00. The results suggest that salinity *per se* does not limit growth of young shrimp.

Zein-Eldin, Z.P. and D.U. Aldrich. *Growth and survival of postlarval Penaeus aztecus under controlled conditions of temperature and salinity.* Biol. Bull. 129(1):199-216. Aug. 1965.

The combined effects of salinity and temperature upon growth and survival of postlarvae of the brown shrimp *Penaeus aztecus*, were studied under controlled conditions.

Zein-Eldin, Z.P. and G.W. Griffith. *The effect of temperature upon the growth of laboratory-held postlarval Penaeus aztecus.* Biol. Bull. 131(1):186-196. August 1966.

The growth of postlarval brown shrimp, *Penaeus aztecus*, was studied in the laboratory at constant temperatures of 15° through 35°C. Growth increased with temperature up to 32.5°C. Maximal increases of growth rate per unit of temperature were observed in the temperature range of 17.5° to 25°C. Survival for one month was markedly decreased at 32.5°C, and no animals survived at 35°. The results suggest that in the laboratory gross production is optimal at temperatures of 22.5° to 30°C. Nonlethal temperatures can have a strong effect on the time required to complete postlarval development.

Zein-Eldin, Z.P. and G.W. Griffith. *An appraisal of the effects of salinity and temperature on growth and survival of postlarval penaeids.* FAO Fish. Rep. 3(57):1015-1026. 1969.

Growth and survival experiments conducted in the laboratory on postlarval *Penaeus aztecus Ives* and *P. setiferus* (Linnaeus) showed that, in general, both species tolerated a broad range of temperature and salinity, but some differences between species existed. The effects of salinity and temperature on growth, survival, and tolerances of postlarvae as observed in the laboratory are discussed as related to overwintering of postlarval *P. aztecus*; rate of growth as observed in nature; time of entry of the two species into the nursery areas; and the effects of simultaneous decrease of salinity and temperature.

9.OYSTERS

Andrews, J.D. *Oyster mortality studies in Virginia V. Epizootiology of MSX; a protistan pathogen of oysters.* Ecology 47(1):19-31. 1966.

MSX, a pathogen of oysters, produced a drastic episootic in high-salinity areas of Chesapeake Bay from 1959 to 1963. The patterns of infection and mortality were determined by imports from disease-free seed-oyster areas.

Andrews, J.D. *Interaction of two diseases of oysters in natural waters.* Proc. Natl. Shellfish. Assoc. (1966) 57:38-49. 1967.

A localized episootic caused by *Dermocystidium marinum* was induced in oysters in the York River, Virginia, to simulate natural epizootics in timing of infections and mortalities.

Andrews, J.D. and J.L. Wood. *Oyster mortality studies in Virginia: VI. History and distribution of Minchinia nelsoni, a pathogen of oysters in Virginia.* Chesapeake Sci. 8(1):1-3. 1967.

An epizootic caused by a sporazoan, *Minchinia nelsoni,* commonly known as "MSX," began in large plantings of oysters at the mouth of the York River, Virginia, in 1959. In 1960 all public and private beds in lower Chesapeake Bay experienced heavy losses. Commercial plantings in this area ceased in May 1960. Three slightly wet years were followed by three very dry years, which permitted the parasite, MSX, to spread farther up Chesapeake estuaries. After seven years, no important changes in patterns of timing or intensity of activity have been observed in epizootic areas. Distribution of MSX in Virginia for typical years is depicted with an attempt to categorize areas as to amount and frequency of damage to be expected. The history of the epizootic and some effects on the oyster industry are described. Prevalences of MSX and death rates over a seven-year period are compared.

Anonymous. *Updated hatchery methods promise oyster boom.* Fishing Gazette, 87(3):22, 47-48, 1970.

The new "off-bottom oyster culture centre" of the Bureau of Commercial Fisheries Laboratory, Oxford, Maryland, is experimenting with many different methods of growing oysters off the bottom.

_____. *Long Island Oysters Farms, Inc.: Artificial rearing of oysters.* 1970.

Oysters are artificially reared by allowing the larvae to set on flexible nylon screens from which they can be periodically removed during the metamorphis stage. As the young oysters grow they are transferred to larger screens.

_____. *Demonstration centre for oyster culture set up in Maryland.* The Irish Skipper No. 73 (February):5, 1970.

In order to encourage the off-bottom culture of oysters, the U.S. Bureau of Commercial Fisheries has set up a special centre at Oxford, Maryland, where the various methods can be demonstrated. There is very little off-bottom culture in the U.S., although there are many beds similar to those in Ireland.

_____. *VIMS improves methods of producing 'cultch-free' spat.* Commercial Fisheries Review, 33(4):22, 1971.

A major obstacle to developing seed oysters in commercial hatcheries at reasonable cost has been the expensive washing and handling of bulky oyster and clam shells used as natural cultch. Scientists of the Va. Institute of Marine Science are now concentrating on improving methods for separating spat from artificial substrate at a very early age—and then growing them in trays and tanks without cultch until they are large enough to be planted on beds.

_____. *Pacific oysters bred from a sea loch.* Fishing News International, 11(7):26-27, 1972.

A project started more than two years ago to breed and raise the large and fast growing Pacific oyster (*Crassostrea gigas*) in the Highlands of Scotland is now moving out of the pilot stage. Its operators are already marketing bulk supplies of seed oyster.

Bahr, L.M. and R.E. Hillman. *Effects of repeated shell damage on gametogenesis in the American oyster, Crassostrea virginica.* Proc. Natl. Shellfish Assoc. (1966) 57:59-62. 1967.

A sample of oysters from an upper Potomax River oyster bar was randomly divided into four groups and held in the laboratory. The shells of groups 1 and 4 were repeatedly filed in order to force the continuous deposition of new shell. Groups 1 and 2 were held in running, unfiltered Patuxent River water. Groups 3 and 4 were held in running water filtered to severely limit their food intake. Histological sections were made from 10 oysters of each group at the completion of the experiment eight months later, and levels of gonad development were compared. The fed oysters revealed considerable gonad maturity compared to the oysters with limited food intake. A less obvious difference was observed between the filed and the undamaged groups, although some retardation in gametogenesis was indicated when filing was accompanied by limited food intake. Both filed groups showed an unusually high ratio of males to females.

Castagna, M., D.S. Haven, and J.B. Whitcomb. *Treatment of shell cultch with Polystream to increase the yield of seed oysters, Crassostrea virginica.* Proc. Natl. Shellfish Assoc. (1968) 59:84-90. 1969.

A commercial-scale study was conducted on the Eastern Shore of Virginia during 1964, 1965, and 1966 to evaluate treatment of shell cultch with Polystream.

Caty, X. *Gill disease in Portuguese oysters on the Atlantic Coast of France. Preliminary note on the presence of proliferations observed on oysters affected by gill disease.* Rev. Trav. Inst. Peches Marit. Paris 33:167-170. 1969.

Gill fragments of *Crassostrea angulata*, cultured in Difco medium, showed proliferations of structures 5-30μ. The length of the filaments was also variable, being inversely proportional to their length. None of these structures showed internal organization. Intermediate forms were seen also. These phenomena occurred in direct ratio to the intensity of the disease. The origin of the structures might be inorganic or organic.

Comps, M. *Gill disease in Portuguese oysters on the Atlantic Coast of France. Observations relating to the gill infection of Portuguese oysters.* Ref. Trav. Inst. Peches Marit. Paris 33:151-160. 1969.

The histological and cytological characteristics of the disease were studied in temporary and permanent slides under the light microscope.

Comps, M. *La maladie des branchies chez les huitres de genre Crassostrea: Caracteristiques et evolution des alterations processus de cicatrisation (Gill disease in oysters of the genus Crassostrea: Characteristics and evolution of alterations; cicatrization processes).* Rev. Trav. Inst. Peches Mar. 34(1):23-44. 1970.

Copeland, B.J. and H.D. Hoese. *Growth and mortality of the American oyster, Crassostrea virginica, in high salinity shallow bays in central Texas.* Publ. Univ. Tex. Inst. Mar. Sci. 11:149-158. 1966.

Mortality of oysters in a polyhaline bay on the central Texas coast was found to be massive.

Couch, J.A. and A. Rosenfield. *Epizootiology of Minchinia costalis and Minchinia nelsoni in oysters introduced into Chincoteague Bay, Virginia.* Proc. Natl. Shellfish Assoc. (1967) 58:51-59. June, 1968.

The introduction of *Crassostrea virginica* into Chincoteague Bay at Franklin City, Virginia, was studied in relation to haplosporidan epizootics from September 1963 to September 1966.

Curtin, L. *Cultivated New Zealand rock oysters.* Fish. Tech. Rep. N.Z. Mar. Dep. 25:1-50. 1968.

This report is written as a beginner's guide for those persons intending to engage in Rock Oyster Farming in New Zealand, and attempts to answer those questions most frequently put by prospective farmers. The basic principles of the cultivation methods are described and are intended as a general guide only, but the method and equipment being used are those adopted by the Marine Department. They can be modified to suit the individual farmer's needs and preferences as he becomes more proficient. This is also the case with all equipment mentioned.

Dahlstrom, W.A. *Survival and growth of the European flat oyster in California.* Proc. Natl. Shellfish Assoc. (1964) 55:9-17, 1967.

A shipment of European oysters from the U.S. Fish and Wildlife Laboratory, Milford, Connecticut, was planted in trays in Tomales Bay, California for studies of survival and growth.

Davis, H.C. *Effects of some pesticides on eggs and larvae of oysters and clams.* Comm. Fish. Rev. 23(12):8-23. 1961.

The effects of several concentrations of 31 compounds, on egg development and survival and growth of bivalve larvae, have been determined.

Davis, H.C. and A. Calabrese. *Survival and growth of larvae of the European oyster at different temperatures.* Biol. Bull. (Woods Hole) 136(3):193-199. 1969.

Drinnan, R.E. *The effect of early fouling of shell surfaces on oyster spatfall.* (Abstr.) Proc. Natl. Shellfish Assoc. (1968) 59:2. 1969.

The attractiveness of cultch materials to setting oysters (*Crassostrea virginica*) is an important factor in both field and hatchery spat collection.

Drinnan, R.E. and J.P. Parkinson. *Progress in Canadian oyster hatchery development.* Can. Fish Cult. 39:3-16. 1967.

The potential for oyster culture and the immediate goals of the oyster hatchery in the Maritime Provinces are outlined. The aims of the Experimental Oyster Hatchery program are twofold. The first is to develop reliable techniques of rearing seed oysters on a commercial scale. The second is to select stocks of oysters with desirable characteristics, such as fast growth and disease resistance. Methods and progress in achieving these goals are summarized.

Dunathan, J.P., R.M. Ingle, and W.K. Havens. *Effects of artificial foods upon oyster fattening with potential commercial applications.* Res. Tech. Ser. Fla. Dep. Nat. Resour. 58:1-39. 1969.

Analyses were made of glycogen accumulation by adult oysters, *Crassostrea virginica*, fed finely ground cornmeal, brown rice, barley, hominy, corn starch, millet, torula yeast, crab meal, whole wheat, cellulose, glucose, aggregated glucose, the alga *Cracilaria sjoestedtii*, and combinations of cornmeal/crab meal, cornmeal/yeast, and cornmeal/brown rice. Determinations of the value of cornmeal fattening under commercial techniques were also made. Cornmeal and rice produced the best results.

Dunnington, E.A. *Survival time of oysters after burial at various temperatures.* Proc. Natl. Shellfish Assoc. (1967) 58:101-103. 1968.

Experimental burials of oysters were made 3 in. deep in containers of soil held in running sea water at five temperature ranges from less than 5°C to over 25°C. Survival time varied from two days in summer to five weeks in winter, showing a direct relationship to temperature.

Engle, J.B. and A. Rosenfield. *Progress in oyster mortality studies.* Proc. Gulf Caribbean Fish Inst. 15:116-124. 1963.

Several mass mortalities in oyster populations have occurred over the past 50 years and ineffectual attempts to control them have been made. A review of some of these is given. Problems that still confront efforts at control are listed as well as research activities in this area within the Bureau of Commercial Fisheries.

Fujiya, M. *Oyster farming in Japan.* Helgolander wiss. Meeresunters. 20:464-479. 1970.

Review of present status of Japan oyster farming.

Gillespie, L., R.M. Ingle, and W.K. Havens. *Glucose nutrition and longevity in oysters.* Quart. Fl. Acad. Sci. 27(4):279-288. 1964.

Studies were conducted to investigate the use of glucose in the nutrition of oysters.

Gras, P. *Gill disease in Portuguese oysters on the Atlantic Coast of France. Research on the organism responsible for gill disease.* Rev. Trav. Inst. Peches Marit. Paris 33:161-164. 1969.

Small parts of oyster *Crassostrea angulata* gills were cultured on maltose, dextrose or thioglycolate media. The organisms found belonged to the normal bacterial flora of oceans and estuaries and no organism was more frequent in infected than in healthy oysters, although there was some evidence that *Dermocystidium marinum* might be implicated.

Hanks, R.W. *Effect of metallic aluminum particles on oysters and clams.* Cheapeake Sci. 6(3):146-149. 1965.

The effect of powdered aluminum, used in photogrammetric studies of water currents on the eastern oyster, *Crassostrea virginica*, and the soft-shell clam, *Mya arenaria*, was studied under laboratory conditions. Although the metal particles were removed by filtering activity of the organisms, no increase in aluminum was detected in the tissues and no increase was observed in water samples.

Haven, D.S. *Supplemental feeding of oysters with starch.* Chesapeake Sci. 6(1):43-51. 1965.

Effects of carbohydrate supplements on shell and tissue (meats) growth of the oyster *Crassostrea virginica* were evaluated by laboratory studies. The study suggests that under estuarine conditions, tissue weights of oysters may be influenced by quantities of starch in planktonic algal cells rather than by the species or volume of plankters ingested.

Haven, D.S. and J. Whitcomb. *Treatment of shell with Polystream to increase survival of oysters (Crassostrea virginica) in Virginia.* Virginia Inst. Mar. Sci. Spec. Sci. Rep. 54:1-9. 1969.

Survival of oysters set on shells treated with Polystream was investigated during 1963 and 1964. Treated and control shells were held in wire bags in the high-salinity intertidal seaside area of the Eastern Shore and in the moderate-salinity subtidal areas of the lower James River. Significantly more spat survived on treated shells than on controls. Differences in survival could not be attributed to absence of drill predation on treated shells.

Hidu, H. *Gregarious setting in the American oyster Crassostrea virginica Gmelin.* Chesapeake Sci. 10:85-94. 1969.

In laboratory cultures of setting oyster larvae, *Crassostrea virginica*, it was noted that few of many exposed cultch shells collected most of the set. Also, setting was very sporadic with respect to time. Several hypotheses were tested to explain such observations. Individual cultch shells were not differentially attractive to setting larvae. However, cultch shells containing 24-hour-old spat or 2-month-old spat attracted more set than unspatted control shells in the same culture and stimulated more set than that received in separate control cultures. Undersides of bottom-most layers of shells attracted more set than the higher layers of shells in laboratory cultures. Two-month-old spat, inside larval-proof plankton mesh bags, stimulated set on shells outside the bags indicating that a water-borne pheromone may be the stimulating agent. These findings confirm the contention of Crisp (1967) that gregarious setting occurs in *C. virginica*.

Hooper, G.H. and D.A. Finkelstone. *Clearing oyster beds of predators.* U.S. Patent 3,498,264 (3rd March 1970).

Starfish and other predators are exterminated from oyster beds by scalding them with jets of superheated steam. The steam is discharged from nozzles placed on the underside of an insulated raft which is hauled over the bed by a surface vessel. The steam does not in any way damage the oysters.

Jones, L. *Oyster production system.* J. Mar. Tech. Soc. 3(4):13-15. 1969.

The Oyster Production System, designed by students at Auburn University, is a fully automated method for continual year-round oyster farming. A harvester-reseeder unit, housing free-swimming spat, moves down stationary belts, the substrate for growing oysters, and the spats are allowed to anchor in depressions in these belts. Internal gaskets on the harvester clear newly attached oysters not in the belt depressions and external doctor blades remove oysters ready for collection. A collection unit shuttles back and forth transporting the mature oysters to a collection area and resupplying the harvester with oyster larvae on its return. Some of the harvested oysters are induced to spawn so that the larvae supply is replenished. The system of 432 cyclically developed belts would increase productivity while decreasing manpower necessary for oyster farming by present methods.

Katansky, S.C., A.K. Sparks, and K.K. Chew. *Distribution and effects of the endoparasite copepod, Mytilicola orientalis, on the Pacific oyster, Crassostrea gigas, on the Pacific Coast.* Proc. Natl. Shellfish Assoc. (1966) 57:50-58. 1967.

Mytilicola orientalis was studied in relation to *Crassostrea gigas* from April 1963 to March 1965 in Humboldt Bay, California; Yaquina Bay, Oregon; Willapa Bay, Oyster Bay and Hood Canal, Washington. No short-term cyclic effects were noted in regard to the incidence or intensity of infestation, although the incidence of infestation at Willapa Bay showed an increasing trend throughout the study. The infested oysters exhibited a lower Condition Index than the non-infested oysters, but there was little evidence of reduction of shell growth in the infested oysters. Survival of the infested oysters was not adversely affected.

Keith, W.J. and H.S. Cochran. *Charting of subtidal oyster beds and experimental planting of seed oysters in South Carolina.* Contrib. Bears Bluff Lab. 48:3-19. 1968.

The experimental transplants have shown that marketable subtidal oysters can be grown from seed in South Carolina waters. This study indicates the necessity of having certain bottom conditions and other factors such as salinity and turbidity for transplanted oysters to effect maximum growth and survival.

Landers, W.S. and E.W. Rhodes. *Some factors influencing predation by the flatworm, Stylochus ellipticus (Girard), on oysters.* Chesapeake Sci. 11(1):55-60. 1970.

Some of the effects of low temperature, low salinity, prey size, and predator source on the predatory activity of *S. ellipticus* on oysters were investigated in the laboratory.

Lasserre, C. *Gill disease in Portuguese oysters on the Atlantic Coast of France. Preliminary results of a histological study on gill disease.* Rev. Trav. Inst. Peches Marit. Paris 33:165-166. 1969.

A histological study was carried out on *Crassostrea angulata*, *C. gigas* and *Ostrea edulis* infected with gill disease. Tissues from the palps, gills and visceral mass were observed in fresh and fixed preparations. Many amoebocytes were found in infected tissues. The visceral mass and genital gland showed no abnormalities. (French)

Linton, T.L., ed. *Proceedings of the oyster culture workshop—July 11-13, 1967.* Atlanta Marine Fisheries Division, Georgia Game and Fish Commission. 1968. 83 p.

The full text of 15 papers covering the biological, ecological, technical and economic aspects of oyster culture are given. Most of the papers concern methodology and feasibility of culturing oysters in the United States.

Longwell, A.C. *The genetic system and breeding potential of the commercial American oyster.* Endeavour 29(107):94-99. 1970.

This review suggests that the greater understanding of the genetics of the oyster, and its utilization in ways analogous to those used in plant breeding, would help the commercial grower. The authors describe work in the cytogenetics and experimental breeding of oysters. The commercial value of specific contributions of genetics will depend on the development of field management practices.

Loosanoff, V.L. *Time and intensity of setting of the oyster Crassostrea virginica in Long Island Sound.* Biol. Bull. 130(2):211-227. 1966.

These observations, conducted in Long Island Sound from 1937 to 1961, dealt with the occurrence and numbers of oyster larvae (*Crassostrea virginica*), and with the time and intensity of their setting.

Lutz, R.A., H. Hidu, and K.G. Drobeck. *Acute temperature increase as a stimulus to setting in the American oyster, Crassostrea virginica (Gmelin).* Proc. Natl. Shellfish. Assoc. (1969) 60:68-71. 1970.

Field work in Delaware Bay has indicated that a rapid increase in temperature such as experienced in the intertidal zone might stimulate setting in *C. virginica*. To test this possibility, "eyed" setting larvae were kept at a constant temperature (approximately 24°C) for eight hours while experimental cultures were at 24°C for four hours, after which the temperature was rapidly increased to 29°C and held for four hours. A significant increase in larval setting rate was apparent with the initial temperature rise and persisted for three subsequent hours. The very high setting rates experienced when all cultures were initiated indicated that factors other than temperature increase may also stimulate setting.

MacKenzie, C.L. *Oyster culture in Long Island Sound 1966-1969.* Comm. Fish. Rev. 32(1):27-40. 1970.

The efficiency of oyster culture in Long Island Sound has increased sharply in the 1966-1969 period. Since the differential between cost and selling price remains wide, production will probably continue to increase. The location and physical condition of oyster beds, sources of seed oysters, growing oysters from seed to market size, significant recent developments, estimates of possible oyster yields, and problems that need to be solved are discussed by the author.

MacKenzie, C.L. *Causes of oyster spat mortality, conditions of oyster setting beds, and recommendations for oyster bed management.* Proc. Natl. Shellfish Assoc. (1969) 60:59-67. 1970.

As part of a study of mortalities of American oyster, *Crassostrea virginica*, the causes of mortality of oyster spat were identified between setting and age of six months and their relative importance compared. The causes were complex and varied widely from bed to bed. They were a result of predation by various species, overgrowth by others, mechanical breakage during transplanting, suffocation by silt, early post-setting mortality (cause unknown), and deaths apparently from starvation and, in one location, from poisoning by a bryozoan. Two factors prevented much larger sets of oysters on commercial beds: a small supply of clean oyster shells available for commercial oyster companies to plant and layers of fouling organisms and silt that accumulate on shells planted for any length of time. Between 1966-68, nevertheless, the total quantity of oysters in Connecticut was greatly increased by additional care given to seed oysters. From the mid-1950s through the mid-1960s the quantity had been no more than 35,000 hectoliters. By the spring of 1968 it had increased to about 350,000 hectoliters. The additional care consisted of overcoming, to a large extent, the three main causes of mortality of seed oysters, namely, predation by starfish, predation by oyster drills and smothering by silt.

Matthiessen, G.C. *Seed oyster production in a salt pond.* Proc. Food-Drugs Sea 1969:87-91. Washington, D.C., Mar. Technol. Soc., 1970.

Ocean Pond Corporation, located on Fishers Island at the eastern end of Long Island Sound, was organized in 1962 for the purpose of supplying seed oysters (*Crassostrea virginica*) to private oyster producers. Since that year production has increased annually to the extent that in the spring of 1969 about 75 million seed oysters were supplied. A description of the physical structure of the pond, and methods of raft culture are given. During 1970 a set failure occurred and hypotheses are advanced to partially explain the failure.

Maurer, D. and K.S. Price. *Holding and spawning Delaware Bay oysters (Crassostrea virginica) out of season: I. Laboratory facilities for retarding spawning.* Proc. Natl. Shellfish Assoc. (1967) 58:71-77. 1968.

Laboratory facilities and techniques for retarding the natural spawning of *C. virginica* from Delaware Bay are described. Oysters were held in the laboratory from 11 May 1967 until September and October 1967 and January 1968 at which times spawning was successfully stimulated. This work demonstrates that oysters do not resorb even when held for periods up to eight months in the laboratory if proper temperatures and quantities of water are provided the brood stocks. The importance of proper conditioning is stressed.

McKee, L.G., and R.W. Nelson. *Culture, handling, and processing of Pacific Coast oysters.* Fish. Leafl. U.S. Fish and Wildlife Serv. 498. 1960. 21 p.

This report gives historical background; method of growing native and Pacific oysters; method of transporting, opening, washing, grading, and packing fresh oysters; method of processing frozen oysters; method of processing canned oysters and oyster products; and general state regulations governing the Pacific oyster industry.

Medcof, J.C. *Trial introduction of European oysters (Ustrea edulis) to Canadian East Coast.* Proc. Natl. Shellfish Assoc. (1959) 50. 1961.

In the spring of 1957, 1958 and 1959, seed oysters beginning their second and third growing seasons were imported from the United Kingdom oyster breeding tanks at Conway, North Wales. They were examined for parasites and extraneous organisms, carefully cleaned and planted in screen-bottomed trays in Passamaquoddy Bay near St. Andrews, New Brunswick. Some were taken in late 1958 to Ellerslie, Prince Edward Island, and held in Malpeque Bay water. Growth was good in Sam Orr Pond, a warm inlet from Passamaquoddy Bay, but poor in the cool open Bay. The oysters brought in by steamer in 1957 suffered a 95 percent mortality within a month after arrival. The 1958 lot was brought in by air freight and showed a post-shipment loss of only 35 percent. The 1959 lot, also air-shipped, suffered only 9 percent loss up to July 15, 1959, but mortalities rose to 53 percent by August 6. Over-winter survival varied greatly and the flagellate *Hexomita* was found in most moribund oysters. It was found in Ellerslie aquarium stock which died after three weeks' exposure to below-zero water temperatures and in St. Andrews aquarium stock which survived reasonably well at water temperatures that remained above 2°C. It was also found in native oysters taken directly from their beds and in a sea scallop which was held in an aquarium with European oysters but not in quahaugs. The oysters in trays under the ice in Sam Orr Pond survived the relatively mild winter of 1957-58 but died during the severe winter of 1958-59 showing heavy *Hexamita* infestation.

Mori, K. *Effect of steroid on oyster. II.* Bull. Jap. Soc. Sci. Fish. 34:997-999. 1968.

The effects of oestradiol-17β on the fertilization and development of *Crassostrea gigas* whose gonads were not yet fully developed was investigated. Oestradiol-3-benzoate (F oestradiol-17β) caused an increase in the rates of fertilization and development and the result supports the suggestion that the metabolism of steroids by oyster sperm might be related to reproduction. Oestradiol-3 benzoate also caused a marked increase in sperm motility but failed to agglutinate spermatozoa, and induced no structural change such as an acrosomal reaction. It is suggested that oestradiol-17β will be practically applicable to the tank breeding of some shellfish in the near future. (Japanese)

Myhre, J.L. and H.H. Haskin. *MSX prevalence in various stocks of laboratory-reared oyster spat.* (Abstr.) Proc. Natl. Shellfish Assoc. (1968) 59:7. 1969.

Minchinia nelsoni (MSX) prevalence and intensity were determined in various 1966 and 1967 year class laboratory-reared stocks of *Crassostrea virginica* spat. Following first exposure to MSX, striking differences in prevalence levels develop among susceptible and resistant stocks.

Provenzano, A.J. *Effects of the flatworm Stylochus ellipticus on oyster spat in two salt water ponds in Massachusetts.* Proc. Natl. Shellfish Assoc. 50:83-88. 1961.

During the summer of 1957 the larvae and juveniles of the flatworm *Stylochus ellipticus* occurred in great abundance in two salt water ponds on Martha's Vineyard Island. Because of the widespread distribution of *S. ellipticus* in many oyster-growing regions in New England and elsewhere the need for a more effective control of its predation on oyster spat is obvious.

Quayle, D.B. *Pacific oyster culture in British Columbia.* Bull. Fish. Res. Board-Can., 196:1-192. 1969.

The anatomy, growth, breeding, productivity and culture of Pacific oyster (*Crassostrea gigas*) in British Columbia is described and the effects of pollution and predators and pests are discussed. Growth is very variable, optimum fatness occurs during April. Successful breeding of the species is irregular except at Pendrell Sound. Spawning occurs but the larvae fail to survive. Productivity is influenced by growth rate, condition and mortality. Mortality is caused by predation, disease, competition for space, silting and cluster separation. Pollution (industrial and bacteriological) is becoming increasingly important. The methods for seeding and harvesting are described. The future of the industry will depend on control of pollution and industrial encroachment.

Ray, S.M. *Cycloheximide: Inhibition of Dermocystidium marinum in laboratory stocks of oysters.* Proc. Natl. Shellfish Assoc. (1965) 56:31-36. 1968.

Cycloheximide (anti-dione), an antifungal antibiotic, prolonged the life of oysters that were naturally infected with *D. marinum*, a lethal fungus parasite. The feasibility of controlling *D. marinum* infections of oysters used for experimentation in closed systems by continuous treatment with a low level (1 µg/ml/wk) of cycloheximide is suggested by the study.

Sayce, C.S. and C.L. Larson. *Willapa oyster studies: Use of the pasture harrow for the cultivation of oysters.* Comm. Fish. Rev. 28(10):21-26. 1966.

The English pasture harrow is used in oyster cultivation to break apart and scatter clusters of oysters. It is also used to prepare oysters for harvest by loosening them from the substrate and removing fouling growth. An area of the Long Island Oyster Reserve, Willapa Bay, Washington, was divided into a control and three lanes to test the effect of the harrow upon Pacific oysters (*Crassostrea gigas*). The control was undragged, lane 1 was dragged once, lane 2 ten times, and lane 3 three times. Condition of samples of oysters was determined for each lane every week during dragging and once each month for six months after completion of the experiment. The experiment showed that Pacific oysters spawned shortly after being dragged while undragged oysters spawned later. Total mortality of oysters dragged 10 times was no higher than that of oysters dragged once only. Dragging oysters once and three times increased Pacific oyster spatfall three and five times, respectively, but dragging more than three times did not increase spatfall further.

Shaw, W.N. *Raft culture of eastern oysters in Chatham, Massachusetts.* Proc. Natl. Shellfish Assoc. (1960) 51:81-92. 1960.

Growth and survival of raft grown oysters was studied in Oyster Pond and Oyster Pond River, Chatham, Massachusetts from 1957 through 1959. The development of raft culture on a self-sustaining basis is a possible solution for saving the declining oyster industry of Massachusetts and for rational utilization of potential oyster resources of the State.

Shaw, W.N. *A fiberglas raft for growing oysters off the bottom.* Prog. Fish. Cult. 22(4):154. 1960.

A fiberglas raft was constructed to support the weight of 25 bushels of marketable oysters during studies of the growth of oysters off the bottom. The raft consisted of a rectangular box with horizontal wings on each long side, each with two rows of holes. Experimental oysters were hung from these holes with nylon and plastic strips. A mooring line was attached at each end, and the total cost was $225. Either the top or bottom can rest in the water, and the raft has proved durable and is a successful method of growing oysters off the bottom.

Shaw, W.N. *Raft culture of oysters in Massachusetts.* Fish. Bull. U.S. Fish and Wildlife Serv. 61(197):481-495. 1962.

The harvest of oysters in Massachusetts has dropped more than 50 percent in the last 50 years. The possibility of growing oysters attached to rafts was tested as a method of culture that might be useful in reviving the declining oyster industry. Oysters suspended from rafts grew about twice as fast as oysters growing on the bottom. Survival of raft oysters was about six times greater than that of bottom grown oysters. This study showed that oysters can reach market size in 2-1/2 years, if they are first suspended from a raft for 14 months. During the final year, raft grown oysters should be placed on the bottom to let the shells thicken. Normally, wild oysters take from four to five years to reach market size in Cape Cod waters. A gross profit of $3.75 per bushel was earned from the raft-grown oysters. This amount compares favorably with the present gross profit of $4.50 per bushel earned by local oystermen who grow oysters on the bottom. This experiment demonstrates that raft culture is commercially feasible in Massachusetts.

Shaw, W.N. *Index of condition and percent solids of raft-grown oysters in Massachusetts.* Proc. Natl. Shellfish Assoc. (1961) 52:47-52. 1963.

Oysters were suspended from a fiberglas raft in Taylors Pond, Chatham, Massachusetts from September 1959 to October 1960. The index of condition showed a low of 8.8 in July 1960, just after spawning, and highs of 14.6 and 14.5 in October 1959 and September 1960 respectively. The average monthly index for the 13-month period was 12.2. The seasonal cycle of percent solids was similar to that reported for oysters in Upper Chesapeake Bay. Oysters must be planted on the bottom at the end of the first year of suspension to make shells thicken and to eliminate losses due to oysters breaking away from strings during the second year of suspension.

Shaw, W.N. *Natural and artificial pond culture of oysters.* Circ. U.S. Fish and Wildlife Serv. 200:24-29. 1964.

Technical and biological problems involved with oyster culture experiments are described. The history of oyster culture in the United States is included. The projects now underway and plans for the future are discussed.

Shaw, W.N. *The growth and mortality of seed oysters, Crassostrea virginica, from Broad Creek, Chesapeake Bay, Maryland, in high- and low-salinity waters.* Proc. Natl. Shellfish Assoc. (1965) 56:59-63. 1965.

Seed oysters from low-salinity water of Broad Creek were transferred and suspended off bottom in the low-salinity water of Tred Avon River and high-salinity water of Chincoteague Bay. Growth and mortality were measured and compared for two years. The rate of shell growth was similar in both areas. A high second-year mortality was observed in Chincoteague Bay which apparently was not caused by high salinity.

Shaw, W.N. *Advances in the off-bottom culture of oysters.* Proc. Gulf and Caribb. Fish. Inst. 19:108-115. 1966.

For the past 10 years the Bureau of Commercial Fisheries has been experimenting with off-bottom culture of oysters, *Crassostrea virginica*, along the east coast of the United States. Earlier studies at Cape Cod, Massachusetts, demonstrated that when oysters are suspended off-bottom growth, survival, and quality are improved. Further studies in Chesapeake Bay have shown that excellent oyster sets can be obtained by suspending shells from rafts. In one area in 1965 more than 20 oyster spat per shell were collected on suspended shells, as compared with five spat per shell for shells on the bottom. Other studies now in progress at Oxford, Maryland, include the off-bottom culture of oysters in natural and man-made ponds. Preliminary findings indicate that natural ponds are excellent for growing and fattening oysters, and artificial ponds can be used to produce seed oysters. On the basis of recent research, the off-bottom culture of oysters appears to have commercial application along the Atlantic Coast.

Shaw, W.N. *Farming oysters in artificial ponds—its problems and possibilities.* (Abstr.) Proc. Natl. Shellfish Assoc. 58:9. 1968.

Four 1/4-acre artificial salt water ponds were put into operation in the fall of 1964 at Oxford, Maryland. Three oyster culture studies were initiated in the ponds: production of seed oysters; comparative growth of seven strains of Chesapeake Bay oysters, and response of oysters to four types of bottom. In addition, a two-year ecological study was made of the invertebrate succession and pond colonization. We believe that artificial ponds have a commercial potential for culturing oysters which will be determined by further biological research.

Shaw, W.N. *Off-bottom culture of oysters in North America.* (Abstr.) Amer. Fish Soc. Annu. Meet. 98:18-19. 1968.

Off-bottom culture of oysters is being practiced either commercially or experimentally in almost every State that borders the Atlantic Coast, Pacific Coast, or Gulf of Mexico. Extensive raft culture is also conducted on the west coast of Canada. Several small companies are rearing oysters from rafts and racks in California and Washington. The future expansion of this method of oyster culture will depend heavily on the results of experiments now being conducted by State and Federal agencies.

Shaw, W.N. *The past and present status of off-bottom oyster culture in North America.* Trans. Amer. Fish. Soc. 98(4):755-761. Oct. 1969.

Off-bottom culture of oysters has attracted great interest by various agencies along the coasts of the United States and Canada. The major commercial interest in off-bottom oyster culture is now centered in the production of seed oysters. The future expansion of off-bottom oyster culture depends heavily on present research by State and Federal agencies.

Shaw, W.N. *New oyster culture center at Oxford, Maryland.* Chesapeake Bay Aff. 2:3. 1969.

An off-bottom oyster culture center will be established this year at the Bureau of Commercial Fisheries Biological Laboratory. The purpose of the center is twofold: (1) to investigate and compare methods of off-bottom oyster culture that may be commercially practicable in the tributaries of Chesapeake Bay; and (2) to demonstrate the variety of methods used to grow oysters in Japan, Korea, Australia.

Shaw, W.N. and G.T. Griffith. *Effects of Polystream and Drillex on oyster setting in Chesapeake Bay and Chincoteague Bay.* Proc. Natl. Shellfish Assoc. (1966) 57:17-23. 1967.

The possibility of increasing the yield of oyster spat, *Crassostrea virginica*, in Chesapeake Bay and Chincoteague Bay by treating shells with Polystream or Drillex (chlorinated benzene mixture where 45 percent is 1,2,3,4-tetrachlorobenzene) was tested during 1963.

Shaw, W.N. and J.A. McCann. *Comparison of growth of four strains of oysters raised in Taylors Pond, Chatham, Massachusetts.* Fish. Bull. U.S. Fish and Wildlife Serv. 63(1):11-17. 1963.

Former buyers of Wareham River, Massachusetts, seed oysters claim that these oysters are slow growing and have a high mortality rate. The purpose of this experiment was to determine whether Wareham River oysters are truly slow growing. The study demonstrates that in a single environment the Wareham River oysters grow slower than oysters from Long Island Sound and Mill Creek. Further studies are necessary to determine the reasons for this apparent slow growth.

Sinderman, C.J. *Oyster mortalities, with particular reference to Chesapeake Bay and the Atlantic Coast of North America.* Rep. U.S. Fish and Wildlife Serv. 569:1-10. 1968.

A number of recent mass mortalities of oysters of the Middle Atlantic States and elsewhere in the world have been attributed to the effects of disease. Man may have aided spread of diseases by transfers and overcrowding of beds. Reduction of this threat to oyster production could be effected by quarantines, development of disease-resistant strains of oysters, and use of environmental barriers (such as low salinity) to the pathogens involved.

Sprague, V., E.A. Dunnington, and E. Drobeck. *Decrease in incidence of Minchinia nelsoni in oysters accompanying reduction of salinity in the laboratory.* Proc. Natl. Shellfish Assoc. (1968) 59:23-26. 1969.

After the discovery of the parasite *Minchinia nelsoni* in 1958, surveys revealed its incidence only in oyster populations where the salinity in the estuarine area was high.

Velez, A. *Experimental cultures of oysters on the east coast of Venezuela.* FAO Fish. Rep. 71.1:159. 1969.

From June 1967 to June 1968 several experiments were carried out regarding the culture of oysters *Crossostria rhizophorae* in two small bays on the northeastern coast of Venezuela. Two types of collectors were used: (1) wooden frames covered with wire, and (2) chains of shells. Once the spat were attached, the collectors were suspended on a floating device anchored 20 yards from shore. The wooden frames, while accumulating larger number of larvae, provided a poor fixing base and overcrowding caused loss of specimens. The shell collectors were more economical, easier to handle and clean, and had an excellent fixation surface.

Walne, P.R. *Breeding of the Chilean oyster (Ostrea chilensis Philippi) in the laboratory.* Nature (London) 197(4868)L676. 1963.

The breeding properties of the Chilean oyster, *Ostrea chilensis*, were compared with those of other oyster species presently being raised in British waters. The morphological changes occurring in the egg and embryo during development in the mantle cavity were described. The Chilean oyster may be well suited to culturing in British waters since the short free-swimming period would reduce loss by tidal current and predation and the lower breeding temperature 13-15°C, would result in earlier spat production.

Walne, P.R. *Observations on the fertility of the oyster, Ostrea edulis.* J. Mar. Biol. Assoc. U. Kingdom 44(2):293-310. 1964.

Estimates were made of the number of larvae held by brooding females collected from the east, south and southwest coasts of England and the Menai Straits. The brood size was related to the size of the oysters when size was measured as diameter, internal volume or dry meat weight, but there was considerable scatter in the data. Part of the scatter was due to variation in condition of the oyster at the time of spawning. The importance of condition was seen both between and within populations. No decline in fertility was found as the breeding season progressed. No loss of larvae during incubation was observed.

Walne, P.R. *Experiments in the large-scale culture of the larvae of Ostrea edulis L.* Fish. Invest. Min. Agr. Fish. Food (Great Britain) Ser. II Salmon Freshwater Fish. 25(4):1-53. 1966.

Batches of 100,000 larvae were reared in polythene bins filled with filtered sea water and enriched with the flagellate *Isochrysis galbana.* A direct relation was found between the growth rate and the proportion of larvae reaching mature size. Experiments tested the effect of: various food densities, plastic and glass containers; the antibiotics penicillin, streptomycin, chloromycetin and auremycin; stirring; oxygen consumption; light, the attraction of oyster extract to metamorphosing larvae are described.

Walne, P.R. *Present problems in the culture of the larvae of Ostrea edulis.* Helgolander wiss Meeresunters. 20:514-525.

Westley, R.E. *Some relationships between Pacific oyster (Crassostrea gigas) condition and the environment.* Proc. Natl. Shellfish Assoc. (1964) 55:19-33. 1967.

A study has been carried out by the Washington Department of Fisheries to learn some of the relationships between the environment and oyster condition. Areas with an adequate supply of nutrients and high sustained phytoplankton production tended to be areas of good oyster condition whereas areas lacking in nutrients and with little phytoplankton production were areas of poor oyster condition. The water movement in the areas was important for creating an optimum environment for phytoplankton production.

Westley, R.E. *Growth and survival of Korean oyster seed in waters of Washington State.* (Abstr.) Proc. Natl. Shellfish Assoc. 1968 (59):13. 1969.

Seven cases of oysters (*Crassostrea gigas*) were sent from the Pusan area of Korea to be tested to determine their suitability for the Washington Oyster Industry. Control oyster seed was from the Mihagi Prefecture in Japan. Identical plantings of the two seed groups were made in two areas: (1) Keyport (an area of known good oyster growth) and (2) Point Witney, an area of known poor oyster growth. After one year, little difference was noted, in either area, in growth, and survival of the oysters. In the second summer, the Miyagi oysters grew at a rate twice that seen in the Korean oysters, although survival in the two areas was comparable in each group. Visual examination indicated differences in the oyster shells of the two groups, the Korean ones having a more fluted appearance. It was assumed that the oyster seed from Korea was of a different race of *C. gigas* than the oysters from Japan.

Wolf, P.H. *Oyster raft cultivation in New South Wales.* Fisherman 3(6):10-13. 1969.

The method of oyster cultivation in New South Wales with the use of rafts is detailed, as it is concluded that the use of this technique to grow oysters from spat size to maturity may be impractical because of the high winter mortality in permanently submerged oysters. Nevertheless, raft culture is considered ideal for achieving rapid growth and prime condition prior to marketing. Heat killing in the summer months can be eliminated by permanent immersion during this period, and a culture method whereby 78 or more traps can be matured in an area 24 feet by 7 feet in a short time merits further research.

Yancey, R.M. *Review of oyster culture in Alaska, 1910-1961.* Proc. Natl. Shellfish Assoc. (1965): 56:65-66. 1968.

There has been limited success with culture of the Pacific oyster, *Crassostrea gigas* in the Coon Cover-Carroll Inlet near Ketchikan since 1910. Since 1938 from 110 to 227 acres of tidelands have been leased yearly for oyster culture in this area. Several companies have participated at different times but the annual harvest has never exceeded 550 gallons of shucked meats.

Yonge, M. *Oyster cultivation.* Underwater Sci. Technol. J. 2(3):138-144. Sept. 70.

Author reviews the history of oyster culture and the problems associated with early methods. Descriptions of edible oysters cultured in various countries are given along with modern methods of culture.

10. MUSSELS

Brenko, M. and H. Calabrese, *The combined effects of salinity and temperature on larvae of the mussel Mytilus edulis.* Mar. Biol. (Berlin) 4(3):224-226. 1969.

The combined effects of salinity and temperature on survival and growth of larvae of the mussel were studied.

Chanley, P. *Larval development of the hooked mussel, Brachidontes recurvus Rafinesque including a literature review of larval characteristics of the Mytilidae.* Proc. Natl. Shellfish Assoc. (1969) 60:86-94. 1970.

B. recurvus larvae were reared from eggs in the laboratory.

Favretto, L. *Commercial and nutritional aspects of the mussel-breeding industry of the Gulf of Trieste.* Bull. Soc. Adriat. Sci., Trieste 56:243-261. 52 ref. 1968.

Havinga, B.H. *Mussel culture.* See Front. 10(3):155-161. 1964.

The techniques and advantages of cultivating the mussel are discussed.

Pauley, G.B. *The pathology of "spongy" disease in freshwater mussels.* (Abstr.) Proc. Natl. Shellfish Assoc. 58:13. 1968.

11. SEAWEED

Bersamin, S.W. and R.B. Banania and R. Rustia. *Protein substitutes for animal feed from seaweed.* FAO Indo-Pacific Fish Council, Tech. Paper 8, 1966.

12. FRESHWATER PRAWNS

Farming giant freshwater prawn. Fish. Newsl. 21(12):23-24. 1962.

Farming the giant Malayan freshwater prawn is discussed. The scientific information necessary for the cultivation of this prawn has been provided by the experiments of Dr. Shao-Wen Hong. The animals' reproductive habits were studied, as well as the best conditions under which eggs would hatch into healthy larvae such as water salinity and type and method of feeding. A method was developed out of this work, and the process simplified and standardized.

Costello, T.J. *Freshwater prawn culture techniques developed.* Amer. Fish. Farmer 2:8-10,27. Jan. 1971.

The author describes the planning, funding and operation for the experimental farming of the freshwater prawn in Florida. Methods have been developed for mass rearing of larvae which can be transplanted easily to remote areas. The building of the first pilot plant has been completed.

Johnson, D.S. *Biology of potentially valuable freshwater prawn with special reference to the riceland prawn Cryphios lanchesteri.* FAO Fish. Rep. 57:233-241. 1968.

C. rosenbergii (de Man) is the only Malaysian prawn which is being commercially exploited. A summary of research on the unexploited prawns is given.

Lewis, J.B. *Preliminary experiments on the rearing of the freshwater shrimp, Macrobrachium carcinus.* Proc. Gulf Caribbean Fish. Inst. 14:199-201. 1962.

M. carcinus is a freshwater shrimp which has been reported from fresh and brackish water in Eastern America from Florida to Southern Brazil. It reaches a large size at maturity, specimens of 200 mm being not unusual.

Ling, S.W. *Studies on the rearing of larvae and juveniles and culturing of adults of Macrobrachium rosengergii.* Curr. Aff. Bull. Indo-Pacific Fish. Coun. 35:1-11. 1962.

Success in rearing larval stages to juveniles under controlled conditions has been accomplished. A simple and practical culturing method is being developed.

Ling, S.W. *Methods of rearing and culturing Macrobrachium rosenbergii de Man.* FAO Fish. Rep. 3(57):607-619. 1969.

Practical methods are given for culturing and farming *M. rosenbergii* on a large scale. Suitable water conditions, food, tanks, ponds, buildings and equipment for all stages are specified. The scheme covers rearing to early juvenile stages under hatchery conditions and the subsequent stocking of ponds and padifields.

Rosalan, S.B., M.N. Delmendo and T.G. Reyes. *Some observations on the biology of the freshwater prawn Macrobrachium lanceifrons, with notes on the fishery.* FAO Fish. Rep. 3(57):923-933. 1969.

This paper reports some observations on the development of *Macrobrachium* spawned in aquaria. Rearing techniques developed for this species should also be applicable to larger species of freshwater prawn. Notes on the shrimp fishery in Laguna de Bay are also presented.

13. TROUT

Anonymous, *Salmon aquaculture deemed feasible.* Amer. Fish Farmer 1:13. June 1970.

A method was developed for the culture of salmonid fishes. Hatchery-reared fry were transferred to nurseries where they became adapted to a salt-water environment. In trial experiments fish raised in salt-water grew from 0.7 to 10 oz. in under six months. When the fish reached fingerling size, they were transplanted to husbandry units consisting of floating pens submerged in a strong tidal current. The water-mix in these coastal sea farms provides an even temperature, an oxygen-rich environment with little waste accumulation and low disease transmissibility. Fish raised on this schedule would reach market size, 8 to 12 oz., during the winter months when fresh salmon are not readily available.

Borell, A.E., and P.M. Scheffer. *Trout in farm and ranch ponds.* Washington. Farmers Bull. U.S. Dept. Agric. 2154. January 1961. [reprinted Oct. 1966.] 17 p.

This bulletin tells how to plan and manage farm and ranch ponds to grow trout for food and recreation and lists some essentials for commercial production.

Bregnballe, F. *Trout culture in Denmark.* Prog. Fish Cult. 25(3):115-120. 1963.

Denmark is the world's largest exporter of pond-reared trout (16.9 million pounds in 1962). Trout are fed fresh salt water fish in earthen ponds—high production per worker. Trout pond diagram and rearing procedures are described. Trout diseases are listed.

Buss, K., D.R. Graff, and E.R. Miller. *Trout culture in vertical units.* Prog. Fish Cult. 32(4):187-191. 1970.

It was found that high yield and good control could be obtained by culturing trout in vertical "silos."

Butterbaugh, G.L. and H. Willoughby. *A feeding guide for brook, brown, and rainbow trout.* Progr. Fish Cult. 29(4):210-215. 1967.

A proved method of determining the amount of food to feed trout is presented. Tables for feeding brook, brown, and rainbow trout are included.

Christensen, N.O. *Trout farming and trout diseases (fungal, viral) in Denmark.* Ann. N.Y. Acad. Sci. 126(1):420-421. 1965.

The author describes the history and methods of trout farming in Denmark. The principal diseases described are: (1) mold infection (Saprolegnia), (2) whirling disease (Myxosoma cerabralis), (3) Furunculosis, and (4) Egtved disease. Water pollution is becoming an important problem.

Cleaver, F. *Recent advances in artificial culture of salmon and steelhead trout of the Columbia River.* Fish. Leaft. U.S. Fish and Wildlife Serv. 623. 1969. 5 p.

The catch of salmon and steelhead trout from fish reared in Program hatcheries increased rapidly beginning in 1965. By 1967 the benefits from operation of these hatcheries appeared to be well in excess of their costs. The Oregon moist pellet diet was the greatest single factor in providing an economically favorable operation. Further advances in hatchery efficiency are expected in the next few years. Conservation agencies believe that the catch of hatchery produced Columbia River fall chinook salmon, coho salmon and steelhead trout can be increased substantially and that the cost per unit of production can be decreased.

Domurat, J. *Zaburzenia wymiany wodnej a tempo wzrostu zarodkow pstraga teczowego (Salmo gairdneri Rick.)* [Disturbances of water exchange and growth rate of rainbow trout embryos]. Zesz. Nauk. Wyzsz. Szk. Rolnic. w Olsztynie 21(4):563-567. Bibliog. 1966. English summary.

The experiments have been carried out on the eggs of rainbow trout, *Salmo gairdneri Rick,* developing in water from fertilization to the closure of blastopore and subsequently transferred to waterless medium (paraffin oil). It has been found out that under these conditions growth of embryos is retarded and increase in the total length of embryos developing in waterless medium amounts only to one third of the length of embryos developing in water. Under these conditions the heart rate is also slowed down and the mortality rate of embryos, developing in waterless medium, is six times greater than the mortality of control embryos.

Eipper, Alfred W. *Effect of hatchery rearing conditions on stream survival of brown trout.* Trans. Amer. Fish. Soc. 92(2):132-139. 1963.

Rearing trout in warmer water may increase their survival after planting in streams having marginally higher summer temps. Survival was higher in 54°F than in 47°F. Acclimation can increase tolerance to high temp. encountered soon afterward (upper incipient lethal temp.).

Groutage, T.M. *Unique Illinois pond supports rainbow trout.* Prog. Fish-Cult. 30(1):9-12. 1968.

Rainbow trout have thrived for over two years in a 1-acre spring fed pond in west central Illinois. This period of survival is unusual, as most Illinois ponds are too warm or lack sufficient dissolved O_2 to support trout. Spawning was observed but no fry were seen.

Haskell, D.C. *Labor to produce fifty tons of trout: A time study.* Prog. Fish Cult. 14(3):87. 1952.

The more efficient use of labor in trout culture is discussed. An overall management plan of operation is suggested, with particular attention to personnel relations and the resolution of employee problems.

Kawamoto, N.Y., S. Fujimura, and M. Tanizaki. *Preliminary report of studies on transportation of live rainbow trout.* Proc. Indo-Pacific Coun. 7(II) 1957.

Physiological and biochemical factors affecting the transportation of live rainbow trout have been studied and it was found that both faeces and urine of the fish tend to increase mortality. A decrease in mortality during transportation may be achieved by removing these excretions and by lowering the water temperature and adding oxygen.

Knight, A.E. *Embryonic and larval development of the rainbow trout.* Trans. Amer. Fish Soc. 93(4):344-355. 1970.

Photographs of development from initial cleavage through hatching of eggs developed in 54°F running spring water.

Liao, Paul B. *Water Requirements for Salmonids.* Prog. Fish. Cult. 33(4):210-215. 1971.

Extensive formulae needed to calculate water flow rates and oxygen conc. rate per lb. of fish (Salmonids) necessary. For salmon, at an average fish weight of 1 lb. at 45°F, you can grow 30 lbs. of fish with a one gpm flow rate, at 70°C, only 5 lbs. of fish can be maintained.

Locke, D.O. and S.P. Linscott. *A new dry diet for landlocked Atlantic salmon and lake trout.* Progr. Fish. Cult. 31(1):3-10. 1969.

Feeding tests to compare the dry Ewos salmon diet and 100 percent beef liver were made in three Maine hatcheries for two years, using landlocked Atlantic salmon and lake trout.

MacCrimmon, H.R. and W.H. Kwain. *Influence of light on early development and meristic characters in the rainbow trout, Salmo gairdneri Richardson.* Can. J. Zool. 47(4):631-637. 1969.

Initial mortality of newly fertilized rainbow trout eggs incubated in artificial light increased with intensity. Further research is imperative if the importance of light as an environmental factor if the early development of fish is to be understood.

Macek, K.J. *Reproduction in brook trout fed sublethal concentrations of DDT.* J. Fish. Res. Board Can. 25(9):1787-1796. 1968.

When sexually maturing yearling brook trout were fed for 156 days with DDT at rates that evidently caused no mortality, fish fed at the lower dosages produced more mature ova than untreated fish. Those fed at the highest dosage produced fewer mature ova than untreated fish. The size of fish (male) at the end of the feeding period tended to increase according to dosage of DDT.

Mason, J.W., O.M. Brynildson, and P.E. Degurse. *Survival of trout fed dry and meat-supplemented dry diets.* Fish-Cult. 28(4):187-192. 1966.

Brook, brown and rainbow trout reared on pellated dry diets from fry to legal size (6 in.) generally survived and grew as well after release into the wild environment as their counterparts reared on dry diets supplemented with fresh fish. There was low overwinter survival of all trout in a stream with constant ice cover during the winter.

McMauley, R.W. and F. Trimborn. *Incubating rainbow trout eggs in heated, recirculated water.* Prog. Fish-Cult. 30(1):64. 1968.

A 300-watt aquarium heater, set in the middle of a 26 gallon capacity trough with 20,000 rainbow trout eggs on a series of trays, kept the temperature some $10°F$ above that of the water supply.

McFadden, T.W. *Effective disinfection of trout eggs to prevent egg transmission of Aeromonas liquefaciens.* J. Fish Res. Board Can. 26(9):2311-2318. 1969.

Treatment of trout eggs with classical disinfectants such as sulfo-merthiolate, merthiolate, and acriflavine proved unreliable for destroying Aeromonas liquefaciens on the eggshell. Viable bacterial cells are carried on the outer surface of the shell only, enabling the use of surface disinfection.

Phillips, A.M. *Salt in trout diets.* Progr. Fish-Cult. 26(2):95. 1964.

A fortified salt such as livestock feeding salt may be used to provide a safety factor in areas of mineral deficient soils.

Phillips, A.M., G.L. Hammer, and E.A. Pyle. *Dry concentrates as complete trout foods.* Prog. Fish-Cult. 26(1):21-24. 1964.

Three pelleted dry foods, fed as complete diets, have maintained brown trout over an 18 month period. Evidence of nutritional disorders, however, precludes the acceptance of these pelleted foods as complete trout diets under production conditions.

Phillips, A.M., G.L. Hammer, J.P. Edwards and H.F. Hosking. *Dry concentrates as complete trout foods for growth and egg production.* Prog. Fish-Cult. 26(4):155-159. 1964.

Over a 24-month period three pelleted foods fed as complete diets maintained brown trout at satisfactory growth rates and with efficient conversions of food into flesh. Mortalities were normal. There was no evidence of anemia. Suggestions are made to correct nutritional disorders that appeared during the experiment. Fish fed one mixture produced excellent eggs whose survival compared favorably with that of eggs spawned from fish fed meat-meal mixtures under similar hatchery conditions. Eggs from fish fed the other two dry mixtures were inferior.

Piper, R.G. *Toxic effects of erythromycin thiocyanate on rainbow trout.* Progr. Fish-Cult. 23(3):134-135. 1961.

Toxic effects were observed in rainbow trout brood stock fed erythromycin thiocyanate for treatment of kidney disease caused by Corynebacterium. The antibiotic was administered in the food at the rate of 4.5 g of drug per 100 lb. of fish per day for 21 days.

Post, G. and M.M. Beck. *Toxicity, tissue residue, and efficacy of Enheptin given orally to rainbow trout for hexamitiasis.* Prog. Fish-Cult. 28(2):83-88. 1966.

Toxicity of Enheptin to rainbow trout was of a low order showing an LD of 390 mg of Enheptin per kilogram of fish per day when given for six days.

Post, G. and R.E. Keiss. *Further laboratory studies on the use of furazolidone for the control of furunculosis of trout.* Progr. Fish-Cult. 24(1):16-21. 1962.

Experiments were made to determine the acute oral toxicity of micronized furazolidone to brown trout and rainbow trout. Force feeding of dosages as high as 500 mg/kg of body weight per day for 14 days did not cause any pathological effort or mortality in either brown trout or rainbow trout. Trials in fish-cultural stations have indicated that furazolidone will control furunculosis in trout. Indications of toxicity have been noted.

Rucker, R.R. and W.T. Yasutake, and G. Wedemeyer. *An obscure disease of rainbow trout.* The Progr. Fish-Cult. 32(1):3-8. 1970.

Dietary deficiency of ascorbic acid causes scoliosis (a degenerative liver disease) and lardosis in rainbow trout; this was aggravated by the pesticide toxaphene (in very small amounts), causing fingerling mortality at Shelton Hatchery, Washington.

Sano, T. *Etiology and histopathology of hexamitiasis and an IPN-like disease of rainbow trout.* J. Tokyo Univ. Fish. 56:23-30. 1970.

In this study, hexamitiasis and the unknown disease were studied with more emphasis to clarify the epizootic characters. In the intestine of hexamitiasis-fish, a lot of the flagellates were found and the infected fish showed the development of catarrhal enteritis which eventually results in the decquamation of the epithelium of the intestine.

Sowards, C.L. *Experiments in hybridizing several species of trout.* Prog. Fish-Cult. 21(4):147-150. 1959.

A good hatch (73.0 percent) was obtained for the F_1 generation of brookinaw trout (brook trout female X lake trout male) in spite of unorthodox handling during the incubation period. There was no significant post-hatching mortality. The fish were sexually mature at the end of four years and proved to have good fertility and high fecundity. Hatching success for two lots of F_2 generation was 56.4 percent and 78.7 percent respectively. Hatching success of a back-cross with a brook trout female was 82.2 percent. A cross between a brookinaw female and a brown trout male yielded a hatch of 4.8 percent. A cross between a brown trout and a brookinaw male produced a hatch of 32.2 percent. Mortality in the following 23 days reduced this to 0.7 percent. Poor results were obtained with the F_1 generation of splake strout—the cross between a lake trout female and a brook trout male. The hatching success of this cross was 38.5 percent.

Swift, D.R. *Activity cycles in the brown trout. I. Fish feeding naturally.* Hydrobiologia 20:241-247. 1962.

The annual and diurnal activity cycle for four naturally feeding brown trout separately confined in netting cages on the bed of Windermere is described. All fish showed a similar annual cycle of maximum activity during May and June one fish showing a second activity during the autumn. The fish also showed a similar diurnal activity rhythm of low activity during the night and increased activity during the day with a pronounced increase at dawn. The possible influence of light and temperature on the fishes' activity is briefly discussed.

Waite, D. and K. Buss. *An automatic feeder for trout.* Progr. Fish-Cult. 25(1):52. 1963.

An automatic mechanical feeder to dispense pellets to trout has proved to be both efficient and economical. The feeder was made from a 5-gallon paint can (hopper) with a cover. A rotor assembly and hopper cone were constructed, with an electric motor (controlled by an automatic time clock) operating the former. The feeder is hung on a cable over the raceway, and can be filled by pulling it to the bank. During the first year of its operation, maintenance and material cost less than the labor of one man.

Westers, H. *Carrying capacity of salmonid hatcheries.* Prog. Fish-Cult. 32(1):43-46. 1970.

An increase in feeding level results in a proportional decrease in carrying capacity. The cube root of the average weight of fish is proportional to the carrying capacity.

An increase in rate of exchange increases the carrying capacity with a value less than proportional to the rate increase.

14. CRUSTACEANS

Avault, J.W., L. Bretonne and E.J. Jaspers. *Culture of the crawfish—Louisiana's crustacean king.* Amer. Fish. Farmer 1:8-14. Sept. 1970.

Crawfish farming in Louisiana is carried out in ricefield, wooded and open ponds with ricefield ponds producing the largest yield. Production ranges from 200-800 lbs./acre, maximum prices rise to $.65 per pound, harvesting becomes uncommercial below $.15 per pound. Best yields are obtained if ponds are flooded as soon as the young are found in burrows.

Ham, B. Glenn. *Crawfish Culture Techniques.* Am. Fish. Farmer 2(5). 1971.

Hanks, R.W. *Chemical control of the green crab, Carcinus maenas.* Proc. Natl. Shellfish Assoc. 52:75-86. 1961.

Development of a method to protect soft-shell clam stocks from green crab predation is considered necessary for efficient management of the New England clam fishery. The development of new organic pesticides has suggested practical and economical methods of controlling this predator.

Idyll, C.P. *Status of commercial culture of crustaceans.* Proc. Food-Drugs Sea 1969:55-64. Washington, D.C., Mar. Biol. Soc. 1970.

This is an extensive review article on the status of crustacean culture throughout the world.

Kensler, C.B. *The potential of lobster culture.* Amer. Fish Farmer 1:8-12,27. Oct. 1970.

The northern lobster, *Homarus americanus*, supports the nation's sixth most valuable fishery industry and is considered the Atlantic Coast's most valuable fishery product. Current investigations point to the suitability of this species for culture purposes due to the relative ease with which it adapts to mating and hatching in captivity and to the favorable effect of increased temperature on the metabolism of the lobster throughout its life cycle.

Reed, P.H. *Culture methods and effects of temperature and salinity on survival and growth of Dungeness crab (Cancer magister) larvae in the laboratory.* J. Fish. Res. Board Can. 26(2):389-397. 1969.

Recent interest in causes of Dungeness crab population fluctuations led to a study of temperature and salinity effects on survival and growth of zoeae.

Rees, G.H. *Progress on blue crab research in the South Atlantic.* Proc. Gulf Caribb. Fish. Conf. (1962) 15:110-115. 1963.

Research on blue crab in the South Atlantic was initiated by the Bureau of Commercial Fisheries on a limited scale in 1957. Tagging of adult crabs has been carried out. Present indications are that there is very little migration between estuaries. Mortality of the zoeal stages does not appear to be directly associated with either salinity or temperature.

Williamson, D.I. *The type of development of prawns as a factor determining suitability for farming.* FAO Fish Rep. 57:77-84. 1968.

The problems of rearing decapod larvae are discussed and it is suggested that selection of species which hatch in the near adult form would largely eliminate some of the problems. Examples of species with direct or rapid development are given. Their distribution and biology are briefly discussed. Other recommendations include studying the chemical composition of the water and its effects on development.

15. MOLLUSKS

Ansell, A.D. *Experiments in mollusk husbandry.* Fish. News. Int., July/Sept. 1964, 3(3):216-219.

Experiments are being conducted at Poole electricity generating station to investigate the factors involved in two stages of the natural cycle: the growth of microscopic planktonic plants in fertilized sea water, and the growth of animals feeding directly on such plants.

Bowbeer, A. *Seed mollusk production.* World Fish. 19(5):44-45. 1970

The latest developments in oyster cultivation as outlined by speakers at the first conference of the Shellfish Association of Great Britain is reported.

Calabrese, A. and H.C. Davis. *Tolerances and requirements of embryos and larvae of bivalve mollusks.* Helgolander wiss. Meeresunters 20, 553-564. 1970.

Davis, H.C. *Shellfish hatcheries present and future.* Annu. Meet. Amer. Fish. Soc. 98:18. 1968.

The need for shellfish hatcheries and the history of their present develop-ment is discussed. Data was given on the food, temperature, salinity, pH and other requirements of larvae of the American oyster and the hard clam as determined at Milford Laboratory and on the susceptibility of these larvae to toxins and pathogens.

Davis, H.C. and R. Ukeles. *Mass culture of phytoplankton as foods for meta-zoans.* Science 134:562-564. 1961.

An apparatus for mass culture of photosynthetic microorganisms has been developed to grow algae for use as foods for larval and juvenile mollusks in studies of their physiological requirements.

Engle, J.B. *The mulluscan shellfish industry current status and trends (oyster, clams, scallops).* Proc. Natl. Shellfish Assoc. (1965) 56:13-21. 1966.

The author sums up current status and trends in the shellfish industry. He reviews and discusses the major crops such as oysters, clams and scallops.

Escarbassiere, R.M. *Status of the biology and culture of mollusks in Venezuela.* FAO Fish. Rep. 71. 1:156. 1969.

Abundant natural populations of several mollusks exist and are being exploited along the extended shoreline of Venezuela. With the concern for rational exploitation of these natural populations there has been research begun.

Glude, J.B. *Criteria of success and failure in the management of shellfisheries.* Trans. Amer. Fish. Soc. 95(3):260-263. 1966.

Three criteria are proposed for evaluating successful management of commercial fisheries: that the resource can be harvested at a profit; that the resource is maintained at a level which will produce the maximum sustained economic yield; that each participant in the fishery is provided the opportunity to obtain an adequate share of the harvest. Variations between criteria for evaluating commercial and sport fisheries are discussed.

_____. *Hard-clam culture method developed at VIMS.* Commercial Fisheries Review, 32(8-9):21. 1970.

A method of protecting hard-clam seed from natural enemies has been devised by scientists of the Virginia Institute of Marine Science.

Landers, W.S. *Infestation of the hard clam: Mercenaria mercenaria, by the boring polychaete worm, Polydora ciliata.* Proc. Natl. Shellfish Assoc. (1966) 57:63-66.

Accidental and experimental infestation of the hard clam by the polychaete worm in the laboratory is described. It seems that, by burying, clams escape attack but exposed clams, especially small ones, can suffer considerable damage to their shells.

Landers, W.S. and E.W. Rhodes, Jr. *Growth of young clams, Mercenaria mercenaria, in tanks of running sea water.* Proc. Natl. Shellfish Assoc. 58:5. 1968.

Loosanoff, V.L. *New shellfish farming.* N. Amer. Wildlife and Nat. Resources Conf. Trans. (29th):332-337. 1964.

The question of which are the most promising areas of mariculture is raised. It is suggested that the propagation of invertebrates such as crabs, shrimps, lobsters, and particularly mollusks is both feasible and highly practical.

Loosanoff, V.L. and H.C. Davis. *Shellfish hatcheries and their future.* Comm. Fish. Rev. 25(1):1-11. Jan. 1963. Also issued as separate 664 of the U.S. Fish and Wildlife Service.

A new era of shellfish farming is heralded. Food, temperature and salinity requirements of bivalve larvae are given briefly. Problems, present status, and prognosis for bivalve hatcheries are considered.

Menzel, R.W. *Shellfish mariculture.* Proc. Gulf Caribb. Fish. Inst. (1961) 14:195-199. 1962.

This paper discusses the techniques for rearing several types of mollusks.

Menzel, R.W. and H.W. Sims. *Experimental farming of hard clams, M. mercenaria, in Florida.* Proc. Natl. Shellfish Assoc. 53, 1963.

_____. *Hybridization in species of Crassostrea.* (Abstr.) Proc. Natl. Shellfish Assoc. 58:6. 1968.

Attempts have been made to hybridize, in all possible combinations, *Crassostrea angulata, C. commercialis, C. gigas, C. iredalei, C. rhizophorea,* and *C. virginica.*

_____. *The possibility of molluscan mariculture in the Caribbean.* FAO Fish. Rep. 71.1:156. 1969.

A description of the mollusk culture in the Caribbean region.

Ohshima, Y. and S. Choe. *On the rearing of young cuttlefish and squid.* [In Japanese.] Bull. Jap. Soc. Sci. Fish. 27(11):979-989. 1961. English summary.

A discussion of the habits of young cuttlefish and squid from how they can be grown to what they eat.

Phibbs, F.D. *Larval rearing of bivalve mollusks.* (Abst.) Proc. Natl. Shellfish Assoc. (1968) 59:12. 1969.

Experiments to determine the optimal temperatures, salinities and density of larvae for laboratory rearing were carried out. The following Pacific coast bivalves were reared from their larval stage through to metamorphosis: Pacific and Kumato oysters, native oyster, European oyster, the cockle clam, butter clam, gaper clam, and razor clam.

Russell, F.S. *Parasites of commercially important marine mollusks.* Adv. Mar. Biol. 5:276-285. 1967.

Six species of gastropods of the family *Pyramidellidae* which are parasitic on commercially important pelecypods are described in detail. Descriptive characteristics, ecology and pathology are given for each species with references to supporting studies.

Ryther, J.H. *The potential of the estuary for shellfish production.* Proc. Natl. Shellfish Assoc. (1968) 59:18-22. 1969.

High yields of shellfish depend upon the concentration of food produced over a large area.

Walne, P.R. *Sea-water supply system in a shellfish-culture laboratory.* Res. Rpt. U.S. Fish and Wildlife Serv. 63:155-159. 1964.

The sea-water supply system of a set of laboratories devoted to the culture of bivalve mollusks is described.

Webber, H.H. and P.P. Riordan. *Molluscan mariculture.* Proc. Gulf Caribbean Fish. Inst. 21:177-185. 1969.

Cultural procedures and the economics of mariculture are detailed and considered in light of the new business opportunities.

16. FISH—GENERAL REFERENCES

Ahlstrom, E.H. *The importance of egg-larval studies in marine fishery management.* Amer. Fish. Soc. Pap. Annu. Meet. (1968) 98:48-49. 1969.

Systematic surveys of fish eggs and larvae have been carried out by agencies participating in the California Cooperative Oceanic Fisheries Investigations (CalCOF) for almost two decades. Surveys have been used to determine the distribution and abundance of spawning populations of widespread pelagic marine fishes—such as the Pacific sardine, *Sardinops caerulea*, and the jack mackerel, *Trachurus symmetricus*. The surveys have been a primary means of assessing the distribution (at time of spawning) and the abundance of latent fishery resources, such as Pacific hake, *Merluccius productus* (now fished in eastern north Pacific by USSR) and Pacific saury, *Cololabis saira* (still unexploited). The surveys are a sensitive means of showing changes in population abundance over time—such as the spectacular increase in abundance of the population of the northern anchovy, *Engraulis mordax*, and the decline in abundance of the Pacific sardine. Surveys also furnish information on abundance of small forage fishes, such as myctophids and bathylagid smelts, which are important in the food-web, but unlikely candidates for commercial exploitation.

Allen, G.H. *A preliminary bibliography of the utilization of sewage in fish culture.* FAO Fish. Circ., 308. 1969. 15 p.

Only papers dealing with the direct utilization of sewage or effluents from sewage treatment plants, or those that appear to have a direct bearing on the subject, have been included. Citations from the year 1894 through 1966 from the world literature are covered.

Allen, J.H. *Fish farm engineering—2.* World Fish. 18(9):42-43. 1969.

In this article the author deals with the problems of water resistance in relation to permeable netting barriers. He describes in detail illustrative types of barriers used in Japan and Scotland. He also discusses natural enclosures and floating enclosures.

Almand, J.D. *Six keys to fishpond management.* Ga. Ext. Serv. Bull. 669. April 1968. 26 p.

This is a publication of the University of Georgia Cooperative Extension Service for the use of fish farm managers. The major areas covered are construction, poisoning, stocking, fertilizing and liming, weed control and proper fishing. Practical suggestions are given for management in each of these areas. Other pond problems such as muskrats, beavers, birds, snakes, turtles are also discussed.

Amlacher, E. *Taschenbush der Fischkrankheiten fur Fischereibiologen, Tierarzte, Fischzuchter and Aquarianer (Handbook of fish diseases for fishery biologists, veterinarians, fish breeders and owners of aquaria).* Jena, Germany, Gustav Fischer Verlag. 1961. 286 p.

This handbook of fish diseases, in German, is the result of several years of theoretical and practical work by the author in the field of fish pathology. It has for its goal the more important aspects of fish diseases presented in a concise and intelligent manner. Through short notes on experimental and diagnostic methods of individual fish diseases, this work is a handy guide for all those interested in the care and economic importance of fish. The work is copiously illustrated and contains a large chart which summarizes the chief symptoms, diagnosis and treatment of the most important fish diseases. Individual fish diseases are described in detail and each disease is treated according to the following scheme: (1) Differential diagnosis and behavior; (2) Experimental techniques; (3) Cause of the disease, morphology and biology of the infectious agent and course of the disease; (4) Pathology, histology and histochemistry of the disease in question; (5) Therapy, prophylaxis and hygiene.

Andrews, J.W., ed. *Proceedings of conference on high density fish culture.* Pub. Skidaway Inst. Oceanog. 1. 1970. 86 p.

Anonymous. *Aquaculture comes to Turkey Point.* Amer. Fish Farmer 1:14-18, 27. July 1970.

The Turkey Point powerplant of the Florida Power and Light Company is providing assistance to the Sea Grant aquaculture research program of the University of Miami. The experimental facilities are being utilized by research personnel in the investigation of techniques in farming of shrimp and pompano. The methods of shrimp farming used in Southeast Asia and Japan are described. A radical departure from the older methods is the attempt to control the entire life cycle of shrimp. The studies at Turkey Point are aimed toward the production of a shrimp brood stock. In addition to the attempt to induce spawning, research is being conducted to find food that will produce adequate growth of shrimp at an economically acceptable cost. Effects of increased water temperature resulting from powerplant discharges is also under investigation.

_____. *Texas marine research station is step toward farming sea.* Commercial Fisheries Review, 32(6):23-24. 1970.

The Texas Parks and Wildlife Department has begun a research program, "to find a way to increase and use the potential protein production in the sea to meet the increasing world demand for food." Present research seeks to determine the importance of ecological factors that affect growth and survival of fish and shellfish; to evaluate mortality by fishing gear and fish tags; and to measure effects of various pollutants on fishery ecology.

Research will also consider the feasibility of cultivating bait shrimp for off-season sale, the artificial propagation and selective breeding of selected species, and development of methods to maintain organisms under conditions conducive to reproduction in artificial habitats.

Researchers are also studying growth and survival of redfish, speckled trout, and southern flounder.

_____. *Russia plans sea farms.* World Fishing. 21(5):52, 1972.

About one-quarter of the world's continental shelf is situated around the USSR (no less than 6.6 million sq. km) and it is not surprising that this nation is intensely interested in using the coastal area to produce food.

Preliminary research by Soviet scientists indicates that there is a rich potential in seaweeds and algae, which has a very rapid growth rate, which can be accelerated by the application of farming techniques. Work in this field is being carried out at the Archangel Marine Algae Laboratory and by the Sakhalin affiliate of the Pacific Marine Fisheries Institute.

_____. *Europe, USSR: Improve culture of freshwater crustaceans and fishes.* Comm. Fish. Rev. 31(6):48. 1969.

A method of breeding has been developed by biologists which has saved crayfish from extinction in Lithuania. Results show up to 90 young can be hatched from fertilized eggs stripped from each female. In spring when water temperatures are between $10°$ and $15°C$, larvae (6-10 days old) are released into ponds. This year, a half million will be released from hatcheries into lakes and ponds in east Lithuania.

_____. *Fish culture in Israel.* World Fishing. 19(4):27-29. 1970.

In Israel, fish culture production is about 9000 tons, or about 40 percent of the total fish production. The main problem is water supply; most if not all the possible sources are already being exploited to the fullest. Utilizing virtually all the water from the sources of the rivers which flowed across the coastal plain to the Mediterranean has quite simply murdered them. The native freshwater fauna have practically disappeared and the rivers flowing through the coastal cities are little better than open sewers. Those less polluted carry water only when there is heavy winter rain.

Fish farming in general is restricted to the coastal plains, where rivers flow at times in winter, the Jordan Valley and the area around Lake Galilee. All the farms have to be so arranged as to drain or pump one set of ponds into another, so that having raised one crop of fish the water can be used again, even several times.

In all there are just under 100 fish farms with a total area of 12,000 acres (4,860 ha). The large size of the farm is due to the exploitation of most of them by kibbutz. Yields based on national averages are of the order of 1470 lb/acre (1650 kg/ha). The species bred are at present restricted to carp, tilapia and grey mullet. There are longer-term ideas of breeding certain sea fish in shallow artificial seawater ponds along the Mediterranean shore.

_____. *Handbook on fish culture in the Indo-Pacific region.* FAO Fish. Biol. Tech. Pap. 14:1-204. 1962.

A few of the Indo-Pacific countries have brought fish culture to high standards, the average yields per unit area of inland waters being much larger than they are in the region as a whole, and the efficiency and effectiveness of the operations are such that fish are available at economic prices. It is the purpose of this bulletin to describe the methods that have been evolved to improve pond culture, and thus to enable other fish farmers, through advisory and extension services, to emulate practices that have proved profitable in similar areas of the region. In doing this, mention is made of suitable species, of some of the limited investigations that have been conducted to obtain more precise information on several aspects of the industry, of fish foods and feeding, selection of breeding stock, diseases and pests, and of various control measures. The Fisheries Division of the Food and Agriculture Organization is assured that the production of edible and acceptable fish can be greatly increased and, when this is done, that the nutrition levels of the people in the region will be raised.

_____. *Progress of fish farming.* British Food Journal, 73(843):112-14, 1971.

Experimental work on artificial fish rearing is going on in many countries. Cultivation of fresh water fish is anything but new. There is an increasing interest in the artificial cultivation of marine fish, including shellfish, with infinitely greater possibilities and also many difficulties.

The greatest success has been with the fast-growing crustaceans. In Japan, shrimps artificially reared are quite extensively marketed.

The idea of hatching fish and feeding them into selected coastal areas to augment natural stocks had to be abandoned early on. From 1965, advances have been rapid. The main findings for this system was that small tanks and cages were best suited to the rearing of fish; more manageable for feeding purposes. The feeding of certain marine species in their larval stages can be satisfactorily carried out with the naupli (larvae) of the bring shrimp. Stock density is important and it has been found possible to maintain plaice in small cages at a density of five fish per square foot (0.093 m^2), with the fish measuring 30 cm (1 ft.) long and weighing up to half a pound (227 g).

Fish reared in artificially heated sea-water grow faster and continuously. Early problems associated with warmed water have been largely overcome, such as removal of chemical wastes from the effluents and control of temperatures; suitable feeding rates have been found from the preliminary trials. Chlorination to prevent settlement of mollusks and growth of slime, has not prejudiced survival or growth rates of fish.

Methods of feeding determine the species of fish and shellfish which can be cultivated. Growth rate varies in different fish; all grow rapidly for a period and then slow down. In sole and plaice the period of rapid growth is about three years, in some other fish it is much longer.

Atz, J.W. and G.E. Pickford. *The use of pituitary hormones in fish culture.* Endeavour 18(71):125-129. 1959.

Fish domestication has been slow as compared with that of insects, birds, and mammals. Chinese had domesticated the carp and goldfish by 1000 A.D. Egg and fry culture led in Europe to the rearing of trout by the 18th century. Failure to reproduce in captivity has led to investigation of the feasibility of injecting pituitary hormones following the discovery by Houssay in Argentina in 1930 of the induction of premature births in small viviparous fish so treated. The history of this effort is traced with the various pituitary products discussed.

Bardach, J.E. *Marine fisheries and fish culture in the Caribbean.* Proc. Gulf and Caribb. Fish. Inst. 10th Annual Session, pp. 132-37. 1957.

Bardach, J.E. *Aquaculture (fish, shell fisheries).* Science (Washington) 161(3846):1098-1106. 1968.

The role of aquaculture in producing high-grade animal proteins for human nutrition is discussed. Raising and tending aquatic animals is mainly practiced in fresh and brackish waters although there are promising pilot experiments and a few commercial applications of true mariculture. Yields vary with the organisms under culture and the intensity of the husbanding care bestowed on them. The products are now mainly luxury foods, but there are some indications that upgrading of the frequently primitive culture methods now in use could lead to increasing yields per unit of effort and to reduced production costs per unit of weight. Under favorable conditions, production of animal flesh from a unit volume of water far exceeds that attained from a unit surface of ground. With high-density stocking of aquatic animals flushing is important, and flowing water or tidal exchange is essential. Combinations of biological and engineering skills are necessary for full exploitation of aquacultural potentials; these are only partially realized because economic incentives may be lacking to tend aquatic organisms rather than to secure them from wild stocks, because of social, cultural, and political constraints. Nevertheless, a substantial development of aquaculture should occur in the next three decades and with it a severalfold increase in total yield.

Bardach, J.E. and J.H. Ryther, eds. *The status and potential of aquaculture. Vol. II. Particularly fish culture.* Washington, D.C., Amer. Inst. of Biological Sciences, 1968. 275 p.

Review deals with the present status of aquaculture in the world with comments on its potential contribution to alleviating the protein shortage in less developed countries. Survey of aquaculture practices and methods in use today includes a cross section of organisms being cultured in countries with different levels of economic development.

Bardach, John E., J.H. Ryther, and W.O. McLarney. *Aquaculture: The Farming and Husbandry of Freshwater and Marine Organisms.* New York: Wiley-Interscience, Pubs. 868 p. 1972.

Extension text covering the major freshwater and marine cultural organisms. Chapters are by family group. The text includes fish, frogs, crustacean, mollusks, seaweeds and freshwater plants.

Berquist, H. and J. Mason. *Automatic feed dispensers.* Progr. Fish-Cult. 28(3):141. 1966.

Several automatic feed dispensers have been tested with lake and rainbow trout over a period of five years. The dispensers are placed at the heads of the tanks and controlled by a time clock. At predetermined intervals, specific quantities of feed are dropped into the tanks. These feeders do not disturb the fish and there is no accumulation of wasted feed on the tank bottoms, making for more efficient indoor propagation. When used outdoors in cold, damp areas, there are problems of condensation and rust with the feeders.

Boonbrahm, M. *Induced spawning by pituitary hormones injection of pond-reared fishes.* Proc. Indo-Pacific Fish Counc. (1968) 13:162-170. 1970.

This paper summarizes the results of the experiments on induced spawning of Pla-sawai *Pangasius pangasius* (Hamilton), Pla-tapien *Puntius gonionotus* (Bleeker), Pla-kamcham *Puntius orphoides* (C. & V.), and exotic Chinese carps *Ctenopharyngodon idella* (Val.), *Hypopthalmichthys molitrix* (C. & V.) and *Aristichthys nobilis* (Richardson). Using pituitary hormones injection method enabled the production of normal fish fry in commercial quantity.

Borisov, P.G. *Fisheries research in Russia: a historical review* (Translated from Russian). Washington, D.C., U.S. Dept. of the Interior, 1964. 187 p.

The history of fisheries research in Russia is recounted from its beginning in the 13th century to the present day. Until around 1928 most of the research involved surveys and expeditions to determine the extent and types of fishes available to fishermen. After this date there was increasing emphasis on fish husbandry and rational management of fishery resources.

Brown, E.E. *The freshwater cultured fish industry of Japan.* Proc. Comm. Fish Farming Conf. (Ga. Univ.) 1:18-21. 1969.

Four major varieties of freshwater fish are cultured: (1) eel, (2) carp, (3) rainbow trout, and (4) ayu or sweetfish. In 1965, amounts cultured were 85 percent of the eel, 72 percent of the common carp, 100 percent of the rainbow trout, and 10 percent of the ayu. The remainder were caught as wild, fresh water fish. The proportions of the fish cultured and the absolute volumes of output have been increasing rapidly. In 1961, a total of 19,874 tons of fish were cultured. Estimates for 1968 indicate an increase to 48,857 tons, or 146 percent, in seven years. There are four different methods of culture used: (1) pond, (2) running water, (3) circulating-filter system, and (4) net culture in lakes. The major difference between the first three of these is the volume of available water used. Net culture means that fish are produced inside net enclosures anchored in large lakes. Production costs vary widely between the different varieties of fish produced and between the four different methods of production. In general, running-water culture resulted in highest absolute volume of production, followed by circulating-filter systems, net culture, and pond culture methods, in that order. The major feed given was pelletized feeds.

Calaprice, T.R. *Genetics and mariculture.* Tech. Rep. Fish. Res. Board Can. 222. 10 p. 1970.

Author indicates ways in which practices employed in management could influence the genetic structure of the resource. It is suggested that the yield of desirable organisms from the ocean can be increased by intervention on the part of man. Levels of intervention include prudent exploitation of present species, introduction of nonindigenous species, modifying the environment, and modifying the species.

Calhoun, John C., Jr. *Food from the sea; a systems approach.* Ocean Industry 3(7):65-68. 1968.

There is a large gap in terms of research and technological development between our food-from-the-sea program and our food-from-the-land system. The former should not be regarded as a simple single commodity system, but as a complex program with many commodity systems. In addition, there is no unique food-from-the-sea method such as FPC or aquaculture, or single critical gap like acquisition hardware. The food-from-the-sea system uses many resource development methods with numerous critical points. Before a systems analysis can be made in a very detailed manner, much more data will have to be acquired.

Christy, Francis T., Jr. *Realities of ocean resources.* J. Mar. Technol. Soc. 3(1):33-38. 1969.

More extensive exploitation of marine resources coupled with the absence of sovereign authority over property rights of the oceans present problems for international cooperation and management. The characteristics and difficulties involved in the management of marine resources, including fisheries, minerals and other uses, are briefly discussed. The author urges that guidelines be established to resolve disputes arising from territorial claims and from access rights to common property resources so that the oceans' wealth will be equitably distributed.

Cole, H.A. *The scientific culture of sea fish and shellfish.* Fish. News. Int. 7(6):20-22, 25-26, 28. 1968.

The cultivation of sea fish and shellfish is discussed and the associated technical and economic problems are examined. Plaice, sole, and oysters have been successfully cultured in Britain in the warm water effluents from nuclear power stations. The cultivation of crustaceans is the next step from the author's viewpoint and plans are now being made to begin this work.

Coon, K.L., A. Larson, and J.E. Ellis. *Mechanized haul seine for use in farm ponds.* Fish. Ind. Res. 4(2):91-108. 1968.

Present methods of harvesting fish from farm ponds are time consuming, laborious, and wasteful of water. This paper supplies information on a mechanized system in which a haul seine and associated equipment are used to capture fish in farm ponds and a conveyor and associated equipment are used to load and weigh the fish into trucks for shipment to market. The mechanized seine works well both in ponds of small or large size and water as deep as 8 ft.

Dexter, R.W. and D.B. McCarraher. *Clam shrimps as pests in fish-rearing ponds.* Progr. Fish-Cult. 29(2):105-107. 1967.

Cyzicus mexicanus is a pest in goldfish rearing ponds in Pennsylvania, and in northern pike rearing ponds in Nebraska. It seriously reduces fish production and increases labor in handling the fish. Preventing the eggs of the shrimp from drying and adding 1 oz of liquid parathion per 10,000 gal of pond water are effective measures in controlling the pest.

Doxtater, G. *Experimental predator-prey relations in small ponds.* Progr. Fish-Cult. 29(2):102-104. 1967.

This study was an attempt to rate various predator fishes as to their pisciferous abilities in suppressing bluegill (*Lempomis macrochirus*) populations in small ponds.

Ellis, J.N. *Hydrodynamic hatching baskets.* Progr. Fish-Cult. 31(2):114-117. 1969.

Need for an open-mesh basket in which *Mugil cephalus* eggs could be hatched or the larvae reared led to the development of the special hatching basket described here. Water introduced at the bottom provides an upwelling current suitable for many uses, especially with pelagic forms.

Elrod, Joseph Harrison. *Dynamics of a fish population subjected to intensive harvesting.* Unpublished thesis, Auburn University, 1966.

This thesis concerns Schnable/Peterson estimates of largemouth bass and bluegill. Catfish: stocking 100 channel catfish eats fingerlings; 1 acre in a standard largemouth bluegill combination is satisfactory for new ponds but catfish do not maintain a population.

Fish, G.R. *The Fish Culture Research Station (Batu Berendam, Malacca, Federation of Malaya).* Proc. Indo-Pacific Coun. 7(II) 1957.

This station has been established to meet a demand for additional scientific data concerning the factors influencing the yield of fish from artificially controlled ponds.

Ford, E.C. *Potentials of pond farm production.* Proc. Comm. Fish Farming Conf. (Ga. Univ.) 1:77-78. 1969.

The author points out that a lucrative market exists for fishery products and be exploited profitably by competent and hardworking businessmen. He emphasizes that fish farming must be approached as any industrial enterprise, and all aspects of a business venture must be considered.

Furukawa, A. *Techniques of cultivation of marine fish, crustaceans and mollusks in low water, particularly artificial cultivation.* Bol. Estud. Pesca 5(1):18-30. 1965.

The cultivation of marine species is one of the most efficient methods of utilizing the rich productivity of fertile interior waters. In Japan the commercial cultivation of various marine species of fish, crustaceans, mollusks, and algae was recently started in parallel with the expansion of conventional types of marine fishing. Tentative efforts are being made to improve the techniques of cultivation and to extend it to additional species on a commercial scale. The current stage of development of cultivation techniques is reviewed for 31 species. (Translated from Japanese.)

Gaudet, J.L. *The status of warm-water pond fish culture in Europe.* FAO Fish. Rep. 2(44):70-87. 1967.

In the European region, fish culture in warm-water ponds goes back to the Middle Ages and, in France and Spain, particularly, was closely tied to religious practices of the time. Although the purpose of fish culture in Europe is often restocking natural waters for sport and commercial fisheries, much intensive fish culture, oriented primarily towards food production is also carried out. Artificial feeding is widely practiced. The food coefficient varies with the size of fish, so advanced methods require the sampling of the weight and rate of growth of the fish regularly. Stocking material usually come from each country's own production, although new species are sometimes imported. In most European countries, governmental assistance is given to fish culturists, either through the establishment of research institutes and experimental stations or in the form of direct financial assistance. In eastern Europe, the governments act as the central planning agencies for fish culture development. The main problems facing warm-water fish culture in Europe are the inadequacy of water supplies, particularly in Poland and Israel, water pollution, and fish diseases in western Europe. A list of species used for warm-water fish culture in Europe is given.

Gitay, A. (Zoology Dept., Univ. of Cape Town, S. Africa). *Marine farming prospects on the South African West Coast.* The South African Shipping News and Fishing Industry Review, (May):50, 51, 53. 1972.

The high primary production along the western coast of South Africa provides a natural resource which is considerably underexploited. Most of the efforts to achieve a higher maritime production are invested in the improvement of fishing methods. The intensification of the fishing activities through sophisticated mechanization often results in overfishing which is specially acute in coastal regions. To restore the balance of nature, conservation measures ought to be introduced. Even if the balance of nature is restored and restrictions are modified, exploitation of the sea through fishing only remains a short-sighted policy. In order to avoid further abuse of this resource, it is necessary to introduce the scientifically based technique of marine fishing.

With the present knowledge of marine biology, integrated cultivation of black sea mussel, rock lobster and, in certain areas, abalone could be introduced.

Goncharov, G.D. *Rubella, a viral fish disease.* Ann. N.Y. Acad. Sci. 126(1):598-600. 1965.

The rubella of fish, infectious abdominal dropsy, hemorrhabic septicemia, and hydropigenous viral neurosis are probably differently named fish diseases which have a common etiology. The clinical picture of rubella is variable in the USSR and it depends on climatic conditions and species of fishes infected. The author discusses the viral and bacterial concepts of rubella and cites work to support each of the theories.

Green, O.L. *Fingerlings to food fish.* Proc. Comm. Fish Farming Conf. (Ga. Univ.) 1:7-12. 1969.

The author discussed methods of rearing fingerlings in ponds, and gives practical suggestions for control of predators, optimum stocking temperatures and conditions, feeding and feeds, and maintenance of environmental quality including weed control. Spawning methods with equipment, selection and care of broodstock and rearing of young are topics discussed in detail.

Hepher, B. *Ten years of research in fishponds fertilization in Israel. The effect of fertilization on fish yields.* Bamidgeh 14(2):29-38. 1962.

Experiments were initiated on a small scale at Sdeh-Nahum station, then continued at Dor, where greater facilities were available. Ponds were fertilized and compared with controls as to yields of fish. Where there was no feeding, the production of fish in the fertilized ponds was far greater than in the controls. Where artificial food was given, there was less of a difference between the fertilized ponds and the controls. Fertilization was accomplished also with the addition of P alone and of N alone. Phosphate was the most effective. The effect of N was less in summer than at other seasons, perhaps because of fixation of N_2 in the summer by blue-green algae, and by the addition of N-rich water that occurs in the summer to maintain constant water levels. Israel waters are naturally rich in K, and effects of the addition of this element have not yet been tested.

Hepher, B. *Ten years of research in fishpond fertilization in Israel. Fertilizers dose and frequency of fertilization.* Bamidgeh 15(4):78-92. 1963.

The amount of phosphorous and nitrogen that can remain dissolved in pond water depends upon the chemical conditions in the water (pH, calcium content, etc.). It is wasteful to add more fertilizers than can be maintained in the water. A standard dose was established for the fishponds in Israel, amounting to 60 kg superphosphate and 60 kg ammonium sulfate per hectare every two weeks for ponds from 80-100 cm deep. Doubling the amount of fertilizer by adding a standard dose every week did not appreciably increase the yield of fish. Also, yield was not increased by adding half standard doses every week. Fertilization with standard doses at the beginning of the season, followed by no further addition of P and N compounds, showed a lower fish yield than in ponds fertilized every two weeks.

Hiatt, R.W. *Food chains and food cycle in Hawaiian fishponds. Part I. The food and feeding habits of mullet (Mugil cephalus), milkfish (Chanos chanos), and the ten-pounder (Elops machnata).* Trans. Amer. Fish. Soc. 74:250-261. 1944.

The food and feeding habits of the three most important market fish in Hawaiian ponds were analyzed to ascertain the position of these fish in the food chains and the food cycle occurring within the ponds.

Hickling, C.F. *Fish culture.* London: Faber and Faber, 1962. 295 p.

The book is concerned with the production of fish for human food, one of the world's most efficient ways of producing protein. Following introductory chapters, the headings include different kinds of fish culture, water, soil, the dry period, pond construction and management, fertilizers, supplementary feeding and foods, fish culture in brackish water, running water, and rice fields, stocking rates and times for best yields, fish genetics and hybridization, fish diseases, and fishponds and public health. The author is former director of the Tropical Fish Culture Research Institute in Malacca. Many examples are cited from Europe, Russia, Israel and the Orient. The book is abundantly illustrated and contains an excellent index and a bibliography with 184 references.

Hickling, C.F. *The farming of fish.* New York, Pergamon Press. 1968. 88 p.

This book considers briefly all the major aspects concerned with fish farming. The chapters deal with the historical significance, the primary source of material and energy in a fishpond, the water supply and pond soil, the basis of fish farming, the biology and stocking of fishponds, and fish farming in the sea. Included is a list of selected readings and a subject index.

Hooper, F.F., H.A. Podaliak, and S.F. Snieszko. *Use of radioisotopes in hydrobiology and fish culture.* Trans. Amer. Fish. Soc. 90(1):49-57. 1961.

Use of radioisotopes has increased rapidly in hydrobiology and fish culture. Their use has been demonstrated and offers opportunities in problems of lake metabolism. Applications and limitations are discussed.

Huet, M. *Treatise on pisciculture.* Brussels, Belgium: Editions Ch. de Wyngaert, 1960. 369 p.

Sections of this book deal with pond construction and management; fish feeding and nutrition (natural and artificial); carp culture; trout culture; culture of pike, coregonids, largemouth bass, Tilapia and eels; pond management, liming, fertilization, weed control, etc.; drainage and recovery of fish from ponds, and problems of sorting, holding and transportation.

Huet, M. and J.A. Timmermans. *Textbook on fish culture—breeding and cultivation of fish.* Fishing News (Books) Ltd., London, 1973. 449 p.

This book describes the construction and layouts of ponds, along with the breeding techniques of salmonids, cyprinids, catfish, eels, and others.

Iversen, E.S. *Mariculture: which way will it go?* Oceans 1(2):80-85. 1969.

Mariculture, i.e., sea farming or marine agriculture, has recently become a familiar term. Many temperate countries including the United States do little sea farming. In contrast, many Asian regions use sea farming to supplement food production. In developed countries, mariculture must be a profit-making venture. The species which show the greatest potential for mariculture are low in the food chain, e.g., oysters. The heavy production of expensive seafoods per acre is considered practical, but the distinction between true sea farming and sea ranching must be kept in mind. Methods for successful mariculture are available, and research programs to provide additional information for these operations are underway.

Job, T.J. *Status of fish culture in the Near East Region.* FAO Fish. Rep. 2(44):54-69. 1967.

The limitations of the inland water regime in this region, which is predominantly aric, delimit its aquacultural development. However, the basins of the major river systems of the Nile, the Euphrates-Tigris and the Indus, and of several small rivers and streams as also a number of lakes, dams, reservoirs, springs, marshes, irrigation canals, and even scattered oasis pools and wet rice fields offer considerable possibilities for augmenting the natural fish crop through transplantation, management and other piscicultural operations. Moreover, recent stimuli provided through the establishment of experiment-cum-demonstration fish farms as in West Pakistan, Sudan, Syria, Iran and the UAR, have resulted in a general awakening in fish culture, which can, in course of time, transform the cultivable waters into productive units to supplement the local protein supply for the growing population. Statistics of pisciculture in the region are meager; nevertheless, available data indicate that with progressive attention of the governments and the people and with the increase in skilled manpower in these countries, fish culture will come to play a role of appreciable importance in the Near East.

Johnson, M.D. *Preliminary experiment on fish culture in brackish-water ponds.* Progr. Fish-Cult. 16(3):131-133. 1954.

An experimental project in brackish-water pond cultures was undertaken at the Marineland Research Laboratory in Florida. It was found that when inorganic fertilizer was added to brackish pond water it did not increase the production of mullet or the other species used enough to justify a commercial pond fishery. Additional experiments are being done on the selection, propagation, and rearing of species that might be produced profitably.

Kennedy, W.A. *Sablefish culture: a preliminary report.* Tech. Rep. Can. Fish. Res. Board 107. 1969. 20 p.

The purpose of this study was to investigate the feasibility of capturing young sable fish and rearing them in captivity to commercial size for profitable sale. The Sablefish Culture project is mainly concerned with the biological aspects of this problem, but the author has based the experimental design of certain economic requirements. Among these requirements are: the proximity of a source of young sablefish, machinery and assured and economical source of fish food, machinery and methods that can be maintained in relatively remote communities, and a source of cheap labor. The results of experiments on tank design, weighing apparatus, foods and feeding, rearing studies, crowding, impoundment and taste are reported. At this point, feeding and adequate nutrients are problems and vitamin supplements are suggested.

Kobayashi, Shinjiro and Hachiro Hirata. *How should the time of food supply be controlled in fish culture?* (Abstr.) Proc. Indo-Pacific Council 7(II). 1957.

Many reports have been published on the correlation between fish growth and kinds of food material. However, no attention has been paid to the time of food supply, although it is easily presumable that fishes have a diurnal rhythm and also annual variation in feeding activity peculiar to each species. Most fish culturists control the time of food supply on the basis of experience or on the basis of their own convenience in management. The authors believe that unless the time of food supply is controlled having concern for both the physiology and behavior of the fish concerned, no rational fish culture can be established. Evidence is presented that the diurnal rhythms of the feeding activity are clearly recognizable in both the goldfish and rainbow trout fry, and that these are heavily distorted by the changes of certain environmental factors.

Laventer, C., Y. Dagan, and D. Mires. *Biological observations in fishponds in the Na'aman region, 1964-1965.* Bamidgeh 20(1):16-30. 1968.

During the years 1960-1963, the yield from fishponds in the Na'aman region was among the lowest in Israel. The average annual yield from the ponds, which represent 14 percent of total national pond area, ranged from 1340-1550 kg/ha as opposed to the national average of 1990-2050 kg/ha. The primary cause for the low yield was due to the low amounts of water (about 20,000 m^3/ha) available to fishponds. The principal water source is that of rain runoff which is drained into the deep ponds during the winter. The pond water level is at its highest in spring, and from July gradually lowers. Due to the slow rate of water warming in spring, carp growth begins late. During the interval between pond filling and the beginning of carp growth, a large crustacean population developed in the pond. Although larvae abound during the August to October period, the lowering of pond water level makes them inaccessible to the fish. The slow heating of the water in spring and the lack of natural food from October onwards necessarily limit the fish-breeding season to 230-250 days. By increasing the fish population density during the first growth season, programming the changeover from the first to the second growth season in June-July, and more efficient use of water, an increase in yields was achieved.

Lewis, D.H., S. McConnell, G.W. Klontz, and L.C. Grumbles. *Infectious diseases and host response of marine fish; a partially annotated bibliography.* College Station, Texas, Texas A&M University Sea Grant Program. October 1970. 104 p.

Lewis, W.M. and R.M. Tarrant. *Sodium cyanide in fish management and culture.* Progr. Fish-Cult. 22(4):177-180. 1960.

Through a deadly poison both as a salt and when converted to hydrogen cyanide gas, sodium cyanide can be safely used by trained persons and is a valuable tool in fish management and culture. The "Cyanegg" is the most convenient form of sodium cyanide for use in fish work. In hatchery or other small ponds, the "eggs" are put in wire cages attached to a line extending across the pond and the cages are then towed back and forth and from side to side over the pond until the cyanide has dissolved. In large and irregular ponds, the "eggs" are placed in a bag and towed behind a boat until dissolved. Sodium cyanide at 0.5 to 1.0 ppm is effective in the live removal of larger centrarchids. It is not, however, very effective for removal of carp and bullheads. Sodium cyanide at 4.0 ppm is the best choice for ridding hatchery ponds of tadpoles, fishes, crayfish, and at least some insects and snails. Large lakes can be sampled with cyanide by closing off a cove with a sheet of polyethylene plastic. To obtain a complete sample, fish that float after the day of poisoning must be picked up. Sampling pool-and-riffle-type streams with sodium cyanide produces a more complete sample than is obtainable by electrical shocking or seining.

Lin, S.Y. and T.P. Chen. *Increase of production in freshwater fishponds by the use of inorganic fertilizers.* FAO Fish. Rep. 44(3):210-225. 1967.

The management of eight ponds with an area of 36.6 ha in a fish farm in Taiwan suffered financial loss for seven consecutive years. A turn from apparently unrecoverable loss to profit was achieved by the use of superphosphate fertilizers in 1965. Heavy stocking of the plankton-feeding silver carp and bighead, in combination with other favorable conditions, contributed to the striking results obtained. The experiment showed that although a heavy dose of 180 kg/ha of P_2O_5 gives a fish yield high enough for profit, the most efficient and economic dose is 40 kg/ha, as fish yield is not proportional to the dosage of P_2O_5. The experiment also revealed that the higher the productivity of a pond, the less P_2O_5 fertilization is needed; perhaps 20 kg/ha may be adequate for maintaining a high level of production. Though the use of complete N-P-K fertilizers entails more expense than the use of superphosphate alone, it does not give a much higher yield. The effect of inorganic fertilizers is limited in ponds where productivity due to mineral nutrients has already reached the maximum.

Lucas, C.E. and B.B. Rae. *Fish farming in temperate waters.* Scott. Fish. Bull. 22:5-9. 1964.

In temperate areas, fish farming is done on a large scale in Japan, the USSR, and several European countries. In the latter countries, most of the farming has been done with the carp and its closely related species, although the rainbow trout has been cultivated in Scandinavian countries. In a number of countries, molluscan shellfish have been cultivated in brackish waters. Some of the basic information and research on fish farming in the United Kingdom is presented, as well as work and plans for future operations in Scotland. The Aberdeen Program, the White Fish Authority, and the possibilities of rearing freshwater fish are considered.

Maar, A., et al. *Fish culture in Central East Africa.* FAO Fish. Rep. #2. 1966.

Excellent source of feasibility for culture of various tilapia, with technological and economic data.

MacKenzie, K. *Parasites in relation to marine fish farming.* Scott. Fish. Bull. 32:21-24. 1969.

Recent developments in marine fish farming have emphasized the importance of studies of parasites effecting marine fish. Parasite-free fish are almost unknown, and in this study a sample of 500 fish from a population of plaice up to two years old yielded 25 different parasite species, 13 of which were present in more than 10 percent of the fish. The most common parasite is the fluke or flatworm. Other parasites are described as well as forms of control such as dilute solutions of formaldehyde. Considerably more research is needed to identify the effects of the parasite on the host and to develop newer control measures.

Matida, Y. *The role of soil in fishpond productivity in Asia and the Far East.* FAO Fish. Rep. 44(3):54-63. 1967.

The paper reviews briefly some roles of soil in the mechanism of fish production in various kinds of fish culture in Asia and the Far East. Supply of nutrients from environs and bottom soil relationship between mineralization of precipitated organic matter and properties of soil, and transformation of nutrient compounds with the change in physiochemical conditions in soil and with various practices such as drying, furrowing and liming are discussed. Attention is drawn to problems which need to be studied further.

McKee, A. *Farming the Sea.* London: Souvenir Press, 1967. 314 p.

McLarney, W.O. *Pesticides and aquaculture.* Amer. Fish Farmer 1:6-7,22-23. 9/70.

Describes hazards of chlorinated hydrocarbons which are not necessarily limited to initial toxic effect on animal or plant life. Known effects include the "multiplying" effect leading to highest concentration at the top of the food chain; metabolites may interfere with the action of steroid sex hormones; tendency to accumulate in fat, liver and brain leads to prevalence in animals containing highest percentage of those tissues. Potential harm in aquaculture is stressed as is restraint in uses of herbicides, suggesting biological control instead.

Meyer, F.P. *Treatment tips. How to determine quantities for chemical treatments in fish farming.* U.S. Bur. Sport. Fish. and Wildlife. RP66. June 1968. 16 p.

Fish farmers often find it difficult to calculate amounts of chemicals for treatment of fish. This pamphlet defines terms, gives calculating procedures, and describes typical situations.

Michael, R.G. *Seasonal trends in physicochemical factors and plankton of a freshwater fishpond and their role in fish culture.* Hydrobiologia. 33(1):144-160. 1969.

A two-year study on seasonal variations of the physicochemical factors and plankton of a freshwater fishpond was made. The importance of these factors in fish culture has been discussed. pH ranging between 7.3 and 8.4 falls within the range considered suitable for fish growth. The amount of dissolved gases at any given time is found to be associated with biological activities taking place in the medium. The relatively high total alkalinity values observed may also help in higher yield of ·fish. Low phosphate content is attributed to utilization by phytoplankton. Most plankters exhibit a single annual peak, though individual species have different seasons of maxima. Volume and dry weight values also showed a single peak suggesting a period of maximum production. However, year-to-year variations observed in total quantity of plankton is likely to affect fish growth and production.

Milne, P.H. *Fish farm enclosures. 1. Marine growths on netting.* World Fish. 18(12):26-28. 1969.

The construction and engineering of fish farm enclosures is detailed. Sites suitable for enclosures are: shallow water (3-7 meters), mid-water (5-12 meters), and deep water (10-12 meters). The best fabric for netting depends on the physical properties of the fabric and its resistance to fouling and deterioration in a marine environment. Some of the fabrics which can be used are made of hemp, cotton, nylon, Terylace, Ultron, Comlene, polyethylene, nethon, Plastabond, galvanized steel, aluminum, stainless steel, brass, copper, empro-nickel, and nickel. The various problems of marine fouling are considered.

Milne, P.H. *Fish farm enclosures. 2. Pressures on the structure.* World Fish. 19(1):32-33. 1970.

The loads and forces which fish farm enclosures have to withstand are described. A table of the breaking load of synthetic and metallic fibers is given, and the problems of severe corrosion, design forces, wind forces, gust speed, and wind forces on netting for various wind speeds are considered. Experimental work with the various types of fibers is reviewed, and the author suggests that for design purposes a figure of 50 percent of the breaking load should be used in calculations to present too much extension.

Milne, P.H. *Fish farm enclosures. 3. Wave forces and currents.* World Fish. 19(2):30-31. 1970.

Methods of calculating the pressures of waves and currents in fish farm enclosures are given. These structures are designed to withstand a "highest" single wave in regard to the average wave heights occurring with a small probability. The design wave height is calculated from the characteristics of the wind field, its direction and speed, fetch length, and water depth variations along the fetch. Thus, each side for an enclosure may require a different design, depending on local conditions.

Milne, P.H. *Fish farm enclosures. 4. Design problems and calculations.* World Fish. 19(3):46-48. 1970.

The manner in which fish farm enclosure nets are designed to compensate for wind, wave, and current forces is outlined. The access catwalk, the primary forces impinging on the net structure, the underlying sediment, and examples of designs for various conditions are presented. Calculations are also given to point out optimum net designs for different situations.

Milne, P.H. *Fish farm enclosures. 5. Methods of retention.* World Fish. 19(4):24-26. Bibliogr. 1970.

The author reviews the range of materials suitable for the protection of fishponds. The advantages of various structures such as nets, mud banks, stone pitching, sand, and synthetic materials such as polyethylene, polyvinyl chloride and butyl rubber are evaluated.

Milne, P.H. *Fish farm enclosures. 6. Impermeable barrier construction.* World Fish. 19(5):28-29. 1970.

An impermeable barrier or barrage across an estuary or tidal inlet allows complete control of the environment behind the barrier. To maintain the marine environment requires sluices or pumps for interchange and replenishment of the sea water. Types of barriers used for marine "farming" throughout Scotland and Ireland are described. Various materials and requirements for specific areas and needs are discussed.

Milne, P.H. *Fish and shellfish farming in coastal waters.* London: Fishing News (Books) Ltd., 1972. 208 p.

This book describes design and construction of hatcheries and ponds including associated hydraulic equipment. Sea rafts, pollution, control of predators for all coastal areas suitable for farming are also discussed.

Moe, M.A., R.H. Lewis, and R.M. Ingle. *Pompano mariculture: preliminary data and basic considerations.* Tech. Ser. Fla. State Bd. Conser. 55:1-65. 1968.

Page, J.W. *Stocking densities and water input relationships in running water cultures.* Pub. Skidaway Inst. Oceanog. 1:24-32. 1970.

At SIO research has been initiated on the intensive culture of catfish. This type of culture, sometimes called tank culture or running water culture, is based on the Japanese culture of salmon and trout and is being modified by SIO investigators for catfish. Presented in this paper are the results of experiments involving freshwater input, water recirculation and stocking density.

Pillai, T.F. *Fish farming in the Philippines, Indonesia and Hong Kong.* FAO Fish. Biol. Tech. Paper 18:1-68. 1962.

This is an extensive report on the methods of brackish and freshwater fish farming as practiced in the Philippines, Indonesia and Hong Kong. More importance was assigned to brackish-water farming methods than to freshwater during these studies. The material has been arrayed and presented as a guide and may also serve as a basis for the classification and accumulation of data on fish farming methods. Subjects covered include methods of farming milkfish, grey mullet and freshwater farming. Construction of ponds, feeds and feeding, predators and parasites, harvesting and processing are also covered.

Pillay, T.V.R. *A bibliography of brackish-water fish culture.* FAO Fish. Circ. 21. 1965. 20 p.

ↃPongsuwana, Umpel. *Progress of Ricefield Fishculture in Thailand.* (Abstr.) Proc. Indo-Pacific Counc. 10(II). 1961.

    Reviews the development of fish culture in rice fields in Thailand and describes the results of the experiments carried out in the Central and Northern Regions of Thailand, on single species as well as mixed species fish culture in rice fields.

Pownall, P.C. *Increasing interest in aquaculture.* Aust. Fish. 28(8):3-8. 1969.

    The author describes the types of aquaculture being undertaken or successfully carried out at various places in the world. Organisms being "farmed" are prawns, crabs, abalone, oysters, mussels, carp, mullet, catfish, yellowtail, trout and plaice. Countries engaged include Australia, China, Denmark, India, Japan, France, Great Britain, Malaysia, Pakistan, the Philippines, the Soviet Union, Singapore, Taiwan, Thailand, and the United States. Technical and economic success of these ventures are discussed.

Post, G. *A review of advances in the study of diseases of fish: 1954-64.* Progr. Fish-Cult. 27(1):3-12. 1965.

    Several advances have been made in the study of diseases of fish in the last 10 years. Systemic diseases have been studied. New methods for using old therapeutic agents have been found and a few therapeutic drugs not previously used to treat fish diseases have been evaluated. External parasitic and bacterial diseases have been somewhat neglected. There is a need for a disinfectant which has greater bactericidal activity and yet is less toxic to fish. A more economical, more effective compound for removing external protozoan parasites has not been found. Histopathology, cell and tissue culture, and immunological procedures have received more attention. The development of disease-resistant strains of fish has continued. Isolation, quarantine, and disinfection of the habitat, as well as procedures for diagnosing fish diseases, have received attention. Several schematic procedures have appeared in the literature. Leaflets pertaining to various diseases, with special emphasis on causitive organisms, have been brought up to date.

Prowse, G.A. *The use of fertilizers in fish culture.* Proc. Indo-Pacific Fish Counc. 9th Session, Section II, pp. 13-75. 1962.

Prowse, G.A. *Neglected aspects of fish culture.* Indo-Pacific Fish Counc. Curr. Aff. Bull. 36:1-9. 1963.

    A number of neglected aspects of fish culture are discussed. Attention is drawn to the importance of the digestibility of the algal food components, and the effect of different fertilizers on the nature of the algae is described. Surface algal scums show disadvantages through autoshading and oxygen deficiency. Over-fertilization is shown to have a deleterious effect, leading to lack of oxygen. The pros and cons of natural aquatic macrophytes and supplementary land plants as fish food are discussed, particularly in relation to oxygen balance. The optimum size of a pond is discussed in relation to the "living-space" phenomenon, evaporation, lateral seepage and vertical capillary movement through bunds, and the considerable loss through transpiration of the plants growing on the bunds. The importance of the ratio of surface area to depth in relation to oxygen balance is stressed. Finally the implications of genetic selection and improvement are discussed, including the possibility of selection pressure under adverse conditions.

Prowse, G.A. *A review of the methods of fertilizing warm-water fishponds in Asia and the Far East.* FAO Fish. Rep. 3(44):7-12. 1967.

Traditional methods of fishpond fertilization are reviewed. In China, Hong Kong, Taiwan, Singapore and Malaysia, Chinese carps are raised in conjunction with pigs and ducks. Cut plants and night soil are often added. In India where other carps are cultured, cut plants, oil seed cake and sewage are used as fertilizers. In brackish-water ponds of Taiwan, Philippines and Indonesia, cut weeds, rice bran, seed cake and manure are added as fertilizer and food. Experiments on artificial fertilizers have been carried out in Taiwan, Philippines and India, but have been based on N-K-P mixtures. In Malacca it has been shown that only phosphate fertilizer is needed in ponds, and this method is more economical in labor and costs.

Purdom, C.E. *Gynogenesis—a rapid method for producing inbred lines of fish.* Fish. New Int. 9:29-32. Sept. 1970.

Rabanal, H.R. *Inorganic fertilizers for pond fish culture.* FAO Fish. Rep. 44(3):164-178. 1967.

The first part of the paper attempts to review the fishpond fertilization practices in the Indo-Pacific region. The second part of the paper explains the experiments on the use of inorganic fertilizers in freshwater ponds conducted in the Auburn University Farm Ponds Project, Auburn, Alabama in 1959.

Rabanal, H.R. *Inorganic fertilizers for pond fish culture.* Phillip. J. Fish. 8(1):23-43.1968.

This paper reviews the fishpond fertilization practices in countries in the Indo-Pacific region, and discusses basic information on the use of inorganic fertilizers obtained from experiments carried out by the author in Auburn, Alabama. Cost and effectiveness of organic and inorganic fertilizers are discussed.

Rafail, S.Z. *Further analysis of ration and growth relationship of plaice, Pleuronectes platessa.* J. Fish. Res. Board Can. 26(12):3237-3241. 1969.

Richards of Rockford, Inc., Rockford, Ill. *Floating aerator.* British Patent 1,194,812 (10 June 1970).

A floating aerator for ponds consists of a stainless steel tank filled with polyethylene foam through which a vertical tube passes. Water is sucked into the tube by an axial flow propeller and discharged above the surface as a circular spray. The propeller is driven by an electric motor mounted above the tube. The equipment is designed for ponds and reservoirs where chemical effluent or hot water is discharged.

Roth, H. and W. Nef. *Intensive rearing of restocking fish in round troughs.* Schweiz. Z. Hydrol. 62(1):251-268. 1967.

In 1964-66 experiments were carried out in the fishery stations of Canton Berne with a view to improving the methods for rearing fish destined for restocking rivers and lakes. The fish were reared in round troughs and fed on dry food. Other experiments were made in order to determine the production possibilities and the value of the fish for restocking purposes.

Ryther, J.H. and G.C. Matthiesen. *Aquaculture, its status and potential.* Oceanus 14(4):2-14. Feb. 1969.

While the potential of the open ocean as a source of food for rapidly growing populations is a subject of widely divergent opinions, there is general agreement among experts about the productivity of the inland and coastal waters. Such productivity has been demonstrated in many parts of the world. The author describes the methods and success of many types of aquaculture. He describes the technical, economic, and social constraints upon aquaculture, the problems in culturing various organisms, synthetic food, and the future of aquaculture, particularly the "hunger problem."

Sarig, S. *Research problems of fish culture.* Bamidgeh 12(1):3-5. 1960.

The water shortage which is a critical factor in Israel has caused changes in the technique of fish culture. Measures to exploit the water supply such as the deepening of ponds, storage of water in winter, maintaining ponds the year around and the increase in pond salinity have created research problems which demand immediate solution. The recent research to combat such problems are reviewed with suggestions for future work.

Sarig, S. *Fisheries and fish culture in Israel in 1967.* Bamidgeh 21(1):3-18. 1969.

The fisheries and fish culture statistics for 1967 are reviewed. There was a slight increase of production and a slight decrease of per capita consumption compared to 1966.

Saunders, R.L. and E.B. Henderson. *Growth of Atlantic salmon smolts and post-smolts in relation to salinity, temperature, and diet.* Tech. Rep. Fish. Res. Board. Can. 149. 1969.

Salmon held in freshwater beyond the smolt stage lived and grew well as or better than others in brackish or full strength sea water. During spring and early summer growth was faster in sea water or brackish water than in freshwater and vice versa in late summer, fall and winter. Observations led to the conclusion that the optimum salinity for growth of Atlantic salmon changes seasonably such that their physiology favors life in the sea during spring and summer and life in freshwater during fall and winter. However, it is not obligatory that salmon go to sea to realize their growth potential; those living in brackish water and freshwater grew larger than others in full strength sea water.

Schuwuman, J.J. *A method for the determination of the suitability of coastal regions for the construction of brackish water ponds.* FAO Indo-Pac. Fish. Counc. Tech. Paper 9, 1964.

Outlines the characteristics required of sites for brackish water ponds construction referring to the characteristics between the site and recorded tidal data. Specifies a minimum effective flooding depth of 40 cm.

Shapiro, S. *Food from the sea and inland waters.* U.S.D.A. Yearb. Agric. 1968:20-33.

Some of the work and research programs of fishery scientists in their attempts to more fully understand the nature of the ocean and its living aquatic resources are described. Various steps in the food web are discussed as well as the production of fish and fish products in terms of changing world needs. Laboratory techniques and methods are also mentioned. Changing natural stocks and methods of artificial cultivation are regarded as vital current problems.

Shelbourne, James E. *Rearing marine fish for commercial purposes.* Calif. Coop. Oceanic Fish. Investment Rep. 10:53-60. 1965.

The author is mainly concerned with whether the hatching and release of young fish is a feasible method of increasing the natural population. He cites experiments that have been conducted to show that such hatchery efforts are useful and describes plaice-rearing studies in Britain. He also describes studies now being started in the hatching, rearing, and harvest of plaice in enclosed sea areas enriched with agricultural fertilizers and warmed with power station effluents.

Schultz, F.T. *Genetic potential in aquaculture.* Proc. Food-Drugs Sea 1969:119-134. 74 ref. Washington, D.C., Mar. Technol. Soc. 1970.

This review discussed many techniques that can be employed in aquaculture genetic improvement programs. Included: population genetics studies, selection and breeding techniques, fertilization and sex control mechanisms, cytological and biochemical methods, and irradiation effects. Adaption of the wild organism to the artificial environment is necessary in order to control production. The benefits of domestication of marine population is compared to that of other commonly known plants and animals.

Sindermann, C.J. *The role and control of diseases and parasites in mariculture.* Proc. Food-Drugs Sea 1969:145-173. 47 ref. Washington, D.C., Mar. Tech. Soc. 1970.

The impact of diseases and parasites on mariculture is emphasized in this article. The effects on cultural marine populations and the methods of control are discussed. The author reviews examples from the limited knowledge of diseases in cultured fish populations and examples from more extensive information about cultured shellfish populations. The effects of introducing fish and shellfish from one geographic area to another, effects of environmental pollutants, the factors prompting outbreaks of disease and various control measures are summarized in this paper.

Sindermann, C.J. *Principal diseases of marine fish and shellfish.* New York: Academic Press Inc. 1970. 369 p.

This book summarizes information about diseases of marine fish and shellfish, intended primarily for individuals and organizations with interests in oceanography and marine biology. A review of the principal diseases of fishes caused by microbial organisms, helminths, parasitic crustacea and genetic and environmentally induced abnormalities and diseases of mollusca and crustacea is followed by consideration of disease-caused mortalities, disease problems in mariculture, internal defense mechanisms, the relation of human diseases to those of marine animals and an assessment of the impact of disease on marine populations. A final discussion involves future research on diseases with attention to commercial species and their value as sources of food.

Smith, F.G. *The chemical control of fish diseases and parasites in freshwater ponds.* Proc. Fish Farming Conf. (Ga. Univ.) 1:13-17. 1969.

The author emphasizes the importance of prevention of diseases and discusses five points for the operator to consider in preventing outbreaks. The major disease organisms encountered in fish culture in southern states are described along with preventive measures and treatments.

Sneed, K.E. and H.P. Clemens. *Hormone spawning of warm-water fishes: its biological and practical significance.* Progr. Fish-Cult. 22(3):109-113. 1960.

The use of gonadotrophins to induce spawning has many advantages in the culture of sport, food, bait, and tropical fishes. Also, it can make available large numbers of genetically similar, equal-aged test animals for radiation, cancer, pollution and genetic studies. Although the use of hormone-induced spawning has progressed in the last 25 years, the biological implications suggested by that work have been largely neglected. The literature implies, however, that there is less specificity in the pituitary hormones of fishes and other vertebrates than previously supposed, that there may be a seasonal difference in the biological effect produced by the pituitary of fishes, and that there is little or no qualitative difference in the pituitary hormones of the male and female fish. Confirmation of these generalities is limited by the lack of good bioassay methods for gonadotrophins from fish and other animals.

Stober, O.J. and W.R. Payne. *A method for preparation of pesticide-free fish food from commercial fish food pellets.* Trans. Amer. Fish. Soc. 95(2):212-214. 1966.

A study of the effects of the assimilation of pesticides by bluegills, required a food in which the desired dosage of pesticides could be controlled qualitatively and quantitatively. Analysis of commercially available dry pelleted fish food revealed several chlorinated hydrocarbon insecticides at low parts per billion levels. Methods and procedures were developed to extract pesticide residues from the dry pelleted fish food. Repelletizing and dietary fortification techniques for the pesticide free meal were also developed. Analysis of the dietary components of the reproduced food indicated the protein, fat and fiber were essentially unchanged. Bluegills were maintained for seven months using this food with no mortality occurring that could be directly attributed to dietary deficiencies.

Swift, D. *Fish farm engineering.* World Fish. 18(7):28-30. 1969.

The pilot scale work on fish farming carried out by the White Fish Authority in Scotland has established a number of practical engineering requirements. These are discussed by the author and include spawning tanks, sea enclosures and fish husbandry.

Swift, D.R. *Marine fish farming.* Scott. Fish. Bull. 32:8-11. 1969.

Some of the problems in marine fish farming and the development of methods in these operations are described. Intensive feeding, efficient husbandry, and the control of water temperature have produced good growth rates in plaice. The process by which such fish can be grown at a profit is described in some detail, but further research is needed for the success of future fish farming attempts. One of the essential problems is the holding enclosure; the optimum type for these fish is not yet known, but will probably be simple onshore ponds above ground level.

Swingle, H.S. *Warm-water ponds for panfish.* U.S.D.A. Yearb. Agric. 1967:169-171.

Warm-water ponds are those in which the summer temperatures rise above 80°F in the upper foot of water. The principal fish raised in these ponds are the bluegill and the largemouth bass. The ponds should be stocked as soon as fish can be gotten from hatcheries and fish food is available. The number of fish that can be raised per acre depends upon the richness of the land in the watershed unless fertilizer is used in the pond. In fertilized ponds, a bluegill-bass combination can yield an annual harvest of 150-170 pounds/acre.

Tabb, D.C., W.T. Yang, C.P. Idyll, and E.S. Iversen. *Research in marine aquaculture at the Institute of Marine Sciences, University of Miami.* Trans. Amer. Fish. Soc. 98(4):738-742. October 1969.

The Institute of Marine Sciences, University of Miami has embarked on a research program to show feasibility of culturing shrimp and pompano. The basic initial assumption of the program is that profit must be the underlying stimulus for research and development of a new industry hence the concentration of effort on animals of high market value.

Tal, S. *Current problems of fish culture.* Bamidgeh 15(4):63-66. 1963.

Based upon projections of future fish consumption in Israel, prognosticated by the Israel Joint Planning Center, the future production of fish in ponds is discussed. Even under the most favorable future conditions, any feasible increase in pond area in Israel producing fish at present rates cannot nearly satisfy the predicted future demands. The only alternative is to increase the rate of production. A reevaluation of fishpond research is in order to explain why so little success has been attained in its primary goal of increasing production.

Thomson, J.M. *Brackish-water fish farming.* Austral. Fish. Newsl. 19(11):17-20. 1960.

Composite account of methods found successful in various countries and considered likely to suit conditions in Australia. Construction of ponds, including location, walls, level, depth of water, exchange of water, subsoil and fertilization are discussed. Fish for ponds, obtaining fish fry, rate of stocking, feeding, rate of yield and harvesting are included. Economics and State Regulation.

Tubb, J.A. *Status of fish culture in Asia and the Far East.* FAO Fish Rep. 2(44):45-53. May 1967.

Fish culture in Asia and the Far East is several thousand years old and the greatest developments have been in brackish water culture for milkfish in Java and the Philippines and pond culture of various species of carps in China and India. The only significant recent introduction is of *Tilapia mossambica*, but methodology has been improved by the introduction of the technique of induced spawning in several of the more important carp species. Culture methods include the farming of brackish and freshwater ponds and the cultivation of fish in rice fields. The use of fertilizers and of supplementary feeds is not well developed except in certain cases where intensive culture is conducted. Until recently most of the stock used in fish culture came from wild-caught fry of the various species except for the common and golden carps which have been bred under culture conditions in most of the countries to which they have been introduced.

Van der Lingen, M.I. *Fertilization in warm-water pond fish culture in Africa.* FAO Fish. Rep. 3(44):43-43. Oct. 1967.

Experience in the use and effects of fertilizers in Africa is summarized and the views of workers on the value of different fertilizing methods and their role in fish culture examined. Organic fertilizers appear to be widely used but in some areas much use has been made of inorganic compounds, mainly phosphates with lime as a "conditioner."

Van Duijn, C. *Diseases of fish.* Ed. 2, Springfield, Ill.: Charles C. Thomas, 1967. 309 p.

Revisions appearing in this second edition include coverage of new drug treatments, among them food additions and injection therapy, which have come into general use since the last edition. Diseases caused by improper feeding, poor water conditions, the presence of toxic substances, attacks from predatory creatures, bacteria, viruses and parasites, and by injuries sustained in physical accidents are covered. This comprehensive review of the main causative agents of fish diseases and the drugs and chemicals available for their treatment forms an accurate and reliable reference source for all aquarists and pondkeepers. There is a subject index.

Vzrma, P. and others. *Hydrological factors and the primary production in marine fishponds.* Indian J. Fish. 10(1, sect. A):197-208. July 1967.

The chemical conditions existing in the experimental fishponds near Palk Bay in Mandapam have revealed the lack of several factors conducive for a balanced growth of animal and plant community. Wide fluctuations in salinity, often reaching hypersaline conditions, combined with very low concentration of essential nutrient salts and their lack of regeneration or replenishment are some of the main reasons for the low level of biological productivity. The basic production rate, calculated from light and dark bottle experiments, is found to be very low compared to that of other commercially working ponds or the open sea. Artificial fertilization of the ponds with only superphosphate has helped to raise to some extent the primary production, but not to a sustained level.

Webber, H.H. *Mariculture.* Bioscience 18(10):940-945. 1968.

The need for animal and human food in the future and the potential of the oceans as producers of this food are discussed. The progress that has been made in the culture of mollusks, crustaceans and fishes in many parts of the world is described in detail with production figures and projected yields.

Wells, V. *Aquaculture in Japan.* Aust. Fish 28(8):14-17. 1969.

The extensive work being carried on by Japanese fisheries is described. Work is proceeding along two lines: cultivation followed by release of juveniles back to natural waters, and cultivation to maturity. Culture of prawns, fish, oysters, mussels, abalone, octopus and crabs are discussed. Implications of this work in establishment of aquaculture in Australia and its long-range effects on food production in the Pacific is emphasized.

Williamson, G.R. *Lessons from Japan in fish farming.* Fishing News Intl. 10(5):30-35. 1971.

Lists main aquaculture species during 1970 in Japan, along with the types of food used to feed fish larvaes. Sources of fish eggs are discussed. Marine aquaculture is stressed, as well as transfer to underdeveloped nations.

Wolny, P. *Fertilization of warm-water fishponds in Europe.* FAO Fish. Rep. 3(44):64-81. Oct. 1967.

The first part of the paper outlines the main problems of fertilization of fishponds under European climatic and economic conditions. The change in opinions, from the theory of nitrogenless fertilization which was believed in during the period between world wars, to the modern approach that recommends inter alia nitrogenous fertilizers, is reviewed. The characteristics of Norway saltpetre, ammonium sulphate and liquid ammonia, which are the three commonly used fertilizers in pond culture, are given. Information concerning experiments conducted with urea and acidic ammonium carbonate is also presented. The second part of the paper reviews the use of the more important organic fertilizers.

Wrobel, S. *The role of soils in fish production in ponds.* FAO Fish. Rep. 44(3):153-163. 1967.

The paper is based on different studies made on ponds in Poland and deals with the role of river drainage basins in the formation of pond soils and the influence of drainage basin soils on the chemical composition of the water irrigating the ponds. It is characteristic of the metabolism in ponds that the synthesis of organic matter outbalances its decomposition and consequently organic matter accumulates on the bottom of ponds. The dependence of the migration of important macroelements on the synthesis and decomposition of organic matter on the O_2 conditions in pond water, are emphasized. Treatments of ponds, aimed at the mobilization of nutrients accumulating in pond soils and at the counteraction of "aging" are discussed. Investigation on pond environment and the importance of simultaneous investigations of soils and water in ponds and the synthesis and decomposition of organic matter are stressed.

Yashouv, A. *Mixed fish culture—an ecological approach to increase pond productivity.* FAO Fish. Rep. 44(4):258-273. 1968.

The state of mixed fish culture in Mediterranean countries is reviewed. The practice has reached an advanced stage only in France, Yugoslavia and Israel. Carp is generally used with tench second in France and Yugoslavia and tilapia in Israel. Different approaches to mixed fish culture in different parts of the world are shown to result from the influence of different socioeconomic, climatic and biological factors. In the Far East, small ponds are used, stocked with five or six species of fish with different feeding habits. High average yields of 3000 kg/ha are obtained. In Europe, large ponds are cropped once a year, with a long (three year) rotation. Carp is the dominant fish, with other species being of minor importance. Production relies largely on natural food. In Israel, moderately sized ponds stocked with two sizes of carp and tilapia or grey mullet are intensively managed. The numbers of each species are adjusted in attempts to reduce both inter- and intra-specific competition and to use the pond resources fully. Yields of 3000 kg/ha or more are obtained. The use of experimental ponds at research stations to evaluate the efficiency of mixed culture techniques is proposed.

Economics Bibliography by Species

CARP

Dukhnovski, M.K. 1970. Economic efficiency of breeding herbivorous fishes in ponds (Ekonomicheskaya effektivnost' vyrashchivaniya rastitel' noyadnkh ryb v prudakh). Rybnoe Khozyaistvo 46(11):101-107. (Obtain translated copy on loan from NMFS).

Patro, J.C., Murthy, T.S., and Rao, M.G. Note on the economics of production of fish seed of indian major carps by adopting induced breeding technique, unpublished typed manuscript, Office of the Director of Fisheries, Government of Orissa, Cuttack-India (no date, circa 1974).

Ranadhir, M. and Krishnan, V. Gopala. The profitability of carp fry culture in Howrah and Hooghly districts of West Bengal, Journal of the Inland Fisheries Society of India, Vol. V, December 1973.

YELLOWTAIL

Yamamoto, Tadasu. 1972. Raising young yellowtail in Japan. Economic aspects of fish production, Internatl. Symposium on Fish. Econ., Paris, Nov. 29 to Dec. 2, 1971, O.E.C.D., pp. 249-362.

EEL

Folsom, William B. 1973. Japan's eel fishery. Mar. Fish. Review 35(5,6):41-45.

OYSTERS

Cavanaugh, Carrol. 1974. Luck, management, laws result in Connecticut oyster 'boomlet'. Nat'. Fisherman, 54(12):4-C.

249

Costello, Frank A.; Marsh, Brent L. 1972. System engineering of oyster production. Univ. Del., Coll. Mar. Studies, Dep. Mech. Aero. Eng., pub. no. 2EN066, Sea Grant GH-109, rep. no. DE1-SG-5-72, 55 pp.

Gaarder, Torbjorn; Bjerkan, Paul. 1934. Oysters and oyster culture in Norway. Bergen, John Griegs Boktrykkeri, 96 pp., 1934; transl., 66 pp., 1959, available from Fish Res. Bd., Canada, Bio. Sta., Andrews, N.B.

Gates, John. 1971. The Atlantic oyster industry. Univ. R.I., Dep. Res. Econ., pp. 9-22.

Gunter, Gordon; Demoran, William J. 1971. Mississippi oyster culture. Amer. Fish Farmer World Aquacult. News 2(5):8-12.

Hidu, Herbert. 1969. The feasibility of oyster hatcheries in the Delaware-Chesapeake Bay region. Proceedings, Conf. Artificial Propagation of Commercially Valuable Shellfish, Oysters, Oct. 22-23, 1969, Univ. Del., Coll. Mar. Studies, pp. 111-131.

Linton, Thomas L. 1969. Feasibility study of methods for improving oyster production in Georgia. Ga. Game and Fish Commission, 179 pp. (Obtain from NTIS, COM-72-11239.)

MacKenzie, Clyde L., Jr. 1970. Oyster culture modernization in Long Island Sound. Amer. Fish Farmer World Aquacult. News 1(6):7-10.

MacKenzie, Clyde L. 1970. Oyster culture in Long Island Sound, 1966-1969. Comm. Fish. Review, Jan., 1970, pp. 27-40.

Marsh, B.L.; Morrison, A.W.; Costello, F.A. 1972. Systems engineering of oyster production. Univ. Del., Aero. Eng. Dep., Sea Grant 2-35223, 21 pp.

Marsh, Brent Luther. 1973. Techniques for design of large-scale systems. Ph.D. thesis, Univ. Del., Dep. Mech. Aero. Eng., 223 pp.

Matthiessen, G.C.; Toner, R.C. 1966. Possible methods of improving the shellfish industry of Martha's Vineyard, Duke's County, Massachusetts. Edgartown, Mass., Har. Res. Foundation, 138 pp. (Obtain from NTIS, PB 1973 095.)

Matthiessen, George C. 1970. A review of oyster culture and the oyster industry in North America. Woods Hole Oceanographic Inst., contrib. no. 2528, 52 pp.

May, Edwin B. 1969. Feasibility of off bottom oyster culture in Alabama. Dauphin Island, Alabama Marine Resources Lab., 19 pp. (Obtain from NTIS, COM-72-10255.)

Pesson, L.L. 1974. The coastal fishermen of Louisiana: their characteristics, attitudes, practices, and responsiveness to change. LSU, Cent. Agri. Sci. Rural Dev., Coop. Ext. Serv., Sea Grant 04-3-1518-19, 60 pp.

Rockwood, Charles E.; Mazek, Warren; Colberg, Marshall; Rhodes, Lewis; Menzel, Winston; Jones, Robert. 1973. A management program for the oyster resource in Apalachicola Bay, Florida. Fla. State Univ., Tallahassee, Fla., 350 pp.

Shaw, William N. 1971. Oyster culture research—off bottom growing techniques. Amer. Fish Farmer World Aquacult. News 2(9):16-19 and 21.

Vaughn, Charles L. 1973. National survey of the oyster industry's problems. Nat'l. Mar. Fish. Serv., Market Res. Serv. Div., contract rep. N-043-41-72. 92 pp.

SHRIMP

de la Bretonne, Laurence W., Jr.; Avault, James W., Jr. 1970. Shrimp mariculture methods tested. Amer. Fish Farmer World Aquacult. News 1(12):8-11, 27.

Fujimura, Takuji. 1972. Development of prawn culture industry. Proceedings, Dauai Aquacult. Conf., Hawaii Dep. Planning Econ. and Kauai county office of Econ. Dev., pp. 5-13.

Garino, David P. 1972. Commodities-Shrimp farming attracts new interest as demand outpaces supply, lifts prices. April 13, 1972, Wall Street, p. 18.

Helfrich, Philip; Ball, John; Berger, Andrew; Bienfang, Paul; Cattell, S. Allen; Foster, Maridell; Fredholm, Glen; Gallagher, Brent; Guinther, Eric; Krasnick, George; Rakowcz, Maurice; Valencia, Mark. 1973. The feasibility of brime shrimp production on Christmas Island. Sea Grant Tech. Rep., UNIHI-SEA-GRANT-TR-73-02, 173 pp. Univ. Hawaii, Sea Grant Prog.

Jhingran, V.G.; Gopalkrishnan, V. 1973. Prospects for the development of brackishwater fish and shrimp culture in India. Technical Conf. on Fish. Management and Dev., Feb. 13-23, 1973, Vancouver, Canada, FI:FMD/73/C-1, Jan. 1973, 6 pp.

Mock, C.R. 1973. Shrimp culture in Japan. Marine Fish. Review, 35(3,4):71-74.

Neal, Richard A. 1973. Progress toward farming shrimp in the United States. Marine Fish. Review, 35(3,4):67-70.

Shang, Yun Cheng. 1972. Economic feasibility of fresh water prawn farming in Hawaii. Univ. Hawaii, Econ. Res. Cent., 49 pp.

Thailand, Ministry of Agriculture, Department of Fisheries. 1971. The result of cost and earning survey on shrimp culture in Thailand, 1970.

Yee, William C. 1970. Potential of aquaculture at nuclear energy centers—a systems study. Oakridge Natl. Lab., Oak Ridge, Tenn., ORNL-4488, 78 pp. (Obtain from NTIS).

SEAWEED

Doty, Maxwell S.; Alvarez, Vicente B. 1973. Seaweed farms; A new approach for U.S. industry, Proceeding, 9th Annual Conference, Wash., D.C., Marine Technological Society, Sept. 10-12, 1973, pp. 701-708.

Ffrench, Rudolph A. 1972. The demand for Canadian seaweeds with special reference to Irish moss. Canadian J. Agri. Econ. 20(2):1-6.

Milne, P.H. 1973. Development of fish farms in Japan. Fishing News Int. 12(12):17-20.

Tanonaka, George. 1973. Summary assessment on the industry and market for seaweed product in the world and United States. Natl. Mar. Fish Serv., Northwest Fish. Cent., unpubl. manuscr., pp. 44.

TROUT

Anonymous. 1972. Bob Erkins talks trout marketing. Fish Farming Ind. 3(2,3, and 5):20-24; 31,32, and 34; and 30.

_____. 1972. Georgia researchers look at trout-catfish rotation. Fish Farming Ind. pp. 12-13.

_____. 1974. Lummi Indians market seafarmed salmon, trout items. Frozen Food Age 22 (10):93.

Araji, A.A. 1972. An economic analysis of the Idaho rainbow trout industry. Univ. Idaho, Coll. Forestry, Wildlife and Range Sci., Dep. Agri. Econ., AE Series No. 118, 9 pp.

Arroyo, S. 1973. Chile develops the cultivation of trout in cages. Fish Farm. Int. 1(1):99-104.

Berge, Leidolv; Farstad, Nelvin. 1971. Norwegian pondfish farming. A paper presented at Internatl. Symposium on Fish. Econ., Paris, Nov. 29 to Dec. 3, 1971, O.E.C.D., no. FI/T(71)1//25. 29 pp.

Brown, E. Evan. 1969. The fresh water cultured fish industry of Japan. Proceedings, Fish farming Conf., Athens, Ga., Jan. 27 to 28, 1969. Univ. Ga., coop. Ext. Serv. Inst. Community area Dev., pp. 18-21.

Brown, E. Evan; Hill, T.K.; Chesness, J.L. 1972. Rainbow trout—a new money crop for South Georgia. Ga. Agri. Res., Univ. Ga., Agri. Exp. Sta., Fall, 1972, pp. 10-12.

Collins, Richard A. 1972. Cage culture of trout in warmwater lakes. Amer. Fish Farmer World Aquacult. News 3(7):4-7.

Gooby, Dick. 1971. Idaho trout farmer profits on 10 cent margin. Fish Farm. Ind. 2(3):36-38.

Klontz, George W. 1973. A survey of fish health management in Idaho. Univ. Ida., Forest, Wildlife, and Range Exp. Sta., info. ser. 3, 34 pp.

McGuinness, Fred. 1973. Prairie pot hole trout: hard way to make a buck. Fish Farming Industry.

Mull, Wilbur C.; Fair, Armor John. 1970. Selected aspects of the market demand for rainbow trout in Atlanta and Northeast Georgia. Univ. Ga., Coll. Bus. Adm., 9 pp.

Pritchard, G.I. 1973. Fish farm projects in Canada. Fish Farming Ind. 1(1):112-114.

CATFISH

Adrian, J.L.; McCoy, E.W. 1971. Costs and returns of commercial catfish production in Alabama. Auburn Univ., Agri. Exp. Sta., Bull. 421, 23 pp.

Adrian, J.L.; McCoy, E.W. 1972. Experience and location as factors influencing income from commercial catfish enterprises. Auburn Univ., Agri. Exp. Sta., Bull. 437, 28 pp.

Allen, Kenneth O. 1970. Mariculture: an infant industry with great potential. Fish Farm. Ind. 1(3):14-16.

Allen, Kenneth. 1971. Progress report on caged catfish. Fish Farm. Ind. (January):28-31.

Anonymous. 1973. Catfish farming risky, larger farms more profitable, study shows. Amer. Fish Farm. World Aquacult. News 4(3):9.

Anonymous. 1970. Catfish profit potential: $179 per acre. Fish Farm. Ind. 1(1):12.

Anonymous. 1970. How much profit in pond culture? Fish Farm. Ind. 1(1):32-35 and 44.

Anonymous. 1969. Proceedings, Commercial Fish Farm. Conf., January 27-28, 1969, Athens, Ga., Univ. Ga., Coop. Ext. Serv. and Inst. Community Area Dev., 85 pp.

Anonymous. 1971. Proceedings, First Annual Kerr Foundation Fish Farming Conference. February 26, 1971. Poteau, Okla. Kerr Foundation, 41 pp.

Anonymous. 1972. Proceedings, Second Annual Kerr Foundation Fish Farming Conference, March 10, 1972, Poteau, Okla. Kerr Foundation, 55 pp.

Anonymous. 1971. Producing and marketing catfish in the Tennessee Valley. Conf. Proceedings, June 30-July 1, 1971, Knoxville, Tenn., T.V.A. 96 pp.

Anonymous. 1972. Raceways versus ponds: How they compare costwise, profitwise. Fish Farm. Ind., 3(1):12-13.

Anonymous. 1972. What's ahead for catfish. Fish Farm. Ind. 3(3):10-12.

Anonymous. 1971. The catfish industry—1971: An economist leads a seminar discussion. Amer. Fish Farm. World Aquacult. News 2(4):12-14, 27.

Ayers, James W. 1971. Marketing problems demand production efficiency and sales promotion (The catfish market: problems and promise). Amer. Fish Farm. World Aquacult. News 2(4):10, 16-17.

Bartonek, Frank A. 1972. Catfish farming. Amer. Fish Farm. World Aquacult. News 3(8):4-7. (Reprinted from Farmland News.)

Billy, Thomas, J. 1973. Pond-grown catfish in the United States: present situation and future opportunities. A paper presented to the FAO Technical Conference on Fishery Products, December 4 to 11, 1973, Tokyo. FAO, FII: FP/73/E-33, 13 pp.

Billy, Thomas J. 1969. Processing pond-raised catfish. Proceedings Fish Farm. Conf. Jan. 27-28, 1969, Athens, Ga. Univ. of Ga., Coop. Ext. Serv. Inst. of Community Area Dev., pp. 42-48.

Brown, E. Evan; LaPlante, M.G.; Covey, L.H. 1969. A synopsis of catfish farming. Univ. of Ga., Coll. Agri. Exp. Sta., Res. Bull. 69, 50 pp.

Brown, Robert H. 1971. Georgia catfish production study evaluates technique. Feedstuffs 43(43):57.

Brown, E. Evan; Holemo, Fred J.; Hudson, Horace. 1973. What the Georgia fee fishing survey reveals. Fish Farm. Ind. 4(3):10, 12 and 13.

Collins, Charles M. 1972. Cage culture of channel catfish (1971 experiment). A paper presented to the 4th annual Convention of the Catfish Farmers of America, February 3 to 5, 1972, Dallas, Texas. The Kerr Foundation, Inc., 23 pp.

Davis, James T. and Hughes, Janice S. 1970. Channel Catfish farming in Louisiana. Baton Rouge. La. Wildlife and Fisheries Com., Wildlife Educ. Bull. 98, 48 pp.

Domahue, Hohn R. n.d., circa 1967. The United States catfish market. U.S. Dep. Interior, Bur. of Com. Fish., processed, 11 pp.

Ford, Erwin C. 1969. Potentials of pond form production, Proceedings Commercial Fish Farm. Conf. Jan. 27-29, 1969. Athens, Ga.: Univ. of Georgia, Coop. Ext. Serv. and Inst. of Community Area Dev., pp. 77-78.

Garner, Carrol R.; Hallbrook, W.A. 1972. Catfish production in Southeastern Arkansas: estimated investment requirements, costs and returns, for two sizes of farms. Univ. Ark., Agri. Exp. Sta., Div. Agri., Rep. Ser. 203, 27 pp.

Gray, D. LeFoy. 1970. How to make a success in the fee fishing business. Fish Farm. Ind. 1(1):28-31.

Greenfield, J.E. 1970. Catfish marketing, 1970. Catfish farmer 2(3):37-44.

Greenfield, J.E. 1970. Economic and business dimensions of the catfish farming industry (revised). U.S. Bur. Comm. Fish., Ann Arbor, Mich., 38 pp.

Greenfield, J.E. 1969. Some economic characteristics of pond-raised catfish enterprises, Proceedings, Conf. on Comm. Fish Farm., Jan. 28-29, 1969, Athens, Ga. Univ. of Ga., Coop. Ext. Serv. and Inst. of Community Area Devel., pp. 67-76.

Greenfield, J.E. 1969. Some economic characteristics of pond-raised enterprises (revised). Natl. Mar. Fish. Serv., Econ. Res. Div., unpubl. manuscr. 23, 19 pp.

Greenfield, J.E. 1970. 1970 profile of the catfish market. Fish Farm. Ind. 1(2):18, 19, and 25.

Grizzel, Roy A., Jr. 1970. Market potential for catfish: 200-300 million pounds by 1975. Fish Farm. Ind. 1(1):18-20, 45 and 46.

Grizzel, Roy A., Jr. 1971. SCS survey shows caged catfish culture not pie in the sky. Fish Farm. Ind. 2(3):17-18.

Heffernan, Bernard E. 1974. Five ways to cut costs on a catfish farm. Fish Farm. Ind. 5(1):16-19.

Hiller, Morton M.; Nash, Darrel A. 1969. The development of catfish as a farm crop and an estimation of its economic adaptability to radiation processing. Nat. Mar. Fish. Serv., Econ. Res. Div., Unpubl. manuscript, 135 pp.

Jones, Walter G. 1969. Market alternatives and opportunities for farm catfish. A paper presented to the Fish Farm. Conf., Oct. 1969, Texas A&M Univ., College Station, Texas. U.S. Bur. Comm. Fish., Ann Arbor, processed, 6 pp.

Jones, Walter G. 1969. Market prospects for farm catfish. Proceedings of the Conf. on Comm. Fish Farm., Jan. 27-28, 1969, Athens, Ga. Univ. of Ga., Cent. for Continuing Educ., pp. 49-61.

Kinnear, H.M. 1972. Catfish spawning—trough culture makes a big difference. Amer. Fish Farm. World Aquacult. News 3(6):6-7.

Kirby, Martin. 1972. Catfish vs. poverty. Amer. Fish Farm. World Aquacult. News 4(1):8-9.

Lee, Jasper S. (N.D.) Catfish farming, reference unit. Miss. State Univ., Curric. Coord. Unit for Vocat. Tech. Educ., viii +103 pp.

Lee, Jasper S. 1973. Commercial catfish farming. Danville, Ill., Interstate Printers and Publishers, 263 pp.

Lewis, William M. 1969. Progress report on the feasibility of feeding-out channel catfish in cages. Farm Pond Harvest 3(3):4-8.

Madewell, Carl E. 1971. Economics and related considerations before entering or expanding a commercial catfish or trout farming operation. Paper presented to Fish Farm. Conf., Nov. 4, 1971, Montgomery Bell State Park, Burns, Tenn., Muscle Schoals, Ala.: T.V.A., 20 pp.

Madewell, Carl E.; Ballew, Ralph J. 1972. Historical development of catfish farming. Amer. Fish Farm. World Aquacult. News 3(3):8-11.

Madewell, Carl E.; Carroll, Billy B. 1969. Intensive catfish production and marketing. T.V.A., Rep. F69ACD6, 30 pp.

Mange, Frank A.; Thompson, Russell G. 1969. An application of an investment model to channel catfish farming. Wash., D.C.: Natl. Mar. Fish Serv., Econ. Res. Div., unpubl. manuscr. 2, 40 pp.

McCoy, E.W.; Ruzic, J.E. 1973. Alabama's recreational catfish ponds. Auburn, Ala.: Auburn Univ., Agri. Exp. Sta., Bull. 451, 21 pp.

McCoy, E.W.; Sherling, A.B. 1973. Economic Analysis of the catfish processing industry. Auburn, Ala.: Auburn Univ. Agri. Exp. Sta., Circular, 207 pp.

McCoy, E.W.; Ruzic, J.E. 1973. Raise catfish for fun and profit—or neither. Auburn Univ. Agri. Exp. Sta., Highlights of Agri. Res., 20(1):8.

Mullins, Troy. 1970. Capital requirements for initiating a catfish production enterprise. Amer. Fish Farm. 1(3):12-14.

Nelson, Roy. 1972. The marketing point of view: processing or pay lakes? Catfish Farmer 4(2):13-14.

Oller, Paul. 1971. Catfish marketing. Proceedings, First Annual Kerr Foundation Fish Farm. Conf. (Feb. 26, 1971). Poteau, Oklahoma 94953: Kerr Foundation, Inc., 41 pp.

Rogers, Bruce D.; Madewell, Carl E. 1971. Catfish farming—cost of producing in the Tennessee Valley region. T.V.A., Natl. Fert. Dev. Cent., circular Z-22, 20 pp.

Russell, Jesse R. 1972. Catfish processing—a rising southern industry. U.S. Dept. Agri., Econ. Res. Serv., Agri. Econ., Rep., 33 pp.

Sullivan, Edward G. 1970. The role of the Soil Conservation Service in the catfish industry in Mississippi. Catfish Farmer 2(3):25-26.

U.S. Dept. Agri., Soil Conserv. Serv. 1973. Cost-returns for catfish food production. Unpubl. table (Obtain from Mayo Martin, ext. biologist) Bureau of Sport Fisheries and Wildlife, Fish Farm. Exp. Sta., P.O. Box 860, Stuttgart, AR 72160, 2 pp.

SALMON

Anonymous. 1972. To market, to market, to buy a small salmon. Pacific Northwest SEA, 5(1):3-8 and 12-13.

Bollman, Frank Herbert. 1971. River basin development and the management of anadromous fisheries: an economic analysis of the Columbia River experience. Ph.D. thesis, Univ. Calif. (Berkeley), Dept. Agri. Econ., 793 pp.

Fraser, Jim; Martin, Stephen G. 1972. The economic and biological feasibility study of rearing chinook salmon, chum salmon and Pacific oysters at the Sguaxin Island, Port Gamble and Skokomish reservations. Federal Way, Wash., Small Indian Tribes Organization of Western Washington, Inc., final report, 60 pp. (Obtain from NTIS, COM-73-10110.)

Mahnken, Conrad V.M.; Novothny, Anthony J.; Joyner, Timothy. 1970. Salmon mariculture potential assessed. Amer. Fish Farmer World Aquacult. News 2(1):12-15, and 27.

McNeil, William J. 1973. Ocean ranching of pink and chum salmon. Univ. Alaska, Sea Grant/Mar. Advisory Prog., Alaska Seas and Coasts, 1(2):1, 6-7.

Richards, Jack Arthur. 1969. An economic evaluation of Columbia River anadromous fish programs. Ph.D. thesis, Ore. State Univ., 274 pp.

Richards, Jack A.; Mahnken, Conrad V.M.; Tanonaka, George K. 1972. Evaluation of the commercial feasibility of salmon aquaculture in Puget Sound. Unpubl. preliminary analysis, Natl. Mar. Fish. Serv., NW Fish. Cent., 35 pp.

Roberts, Kenneth J., n.d. Economics of hatchery salmon disposal in Oregon. Ore. St. Univ., Marine Advisory Program, S.G. 7, 20 pp.

Stevens, Joe B.; Mattox, Bruce W. 1973. Augmentation of salmon stocks through artificial propagation: methods and implications. Adam A. Sokoloski, editor, Ocean Fishery Management: Discussion and Research, NOAA, Natl. Mar. Fish. Serv., Tech. Rep., Circ 371, pp. 133-145.

Wagner, Louis C. 1973. An evaluation of the market for pan-sized salmon. Natl. Mar. Fish. Serv., Market Res. Serv. Div., Res. Contract Rep. N208-1344-72N, 50 pp.

Wahle, Roy J.; Vreeland, Robert R.; Lander, Robert H. 1974. Bioeconomic contribution of Columbia River hatchery coho salmon, 1965 and 1966 broods, to the Pacific salmon fisheries. Fishery Bull. 72(1):139-169.

CRAYFISH

Anonymous. 1970. Crawfish: a Louisiana aquaculture crop. Amer. Fish Farmer World Aquacult. News 1(9):12-15.

Anonymous. 1974. Crawfish: Citgo gets a lot of mileage from unique Louisiana farm. Catfish Farmer World Aquacult. News 6(3):10-12.

Avault, James W., Jr.; de la Bretonne, Larry W.; Jaspers, Edmonde J. 1970. Culture of the crawfish, Louisiana's crustacean king. Amer. Fish Farmer, World Aquacult. News. 1(10):8-14, 27.

Blades, Holland C., Jr. 1974. The marketing distribution of South Louisiana crayfish. A paper presented at the Second International Crayfish Symposium, Baton Rouge, La., April 1974, (pages unknown).

Gary, Don L. 1974. The commercial crawfish industry of South Louisiana. La. State Univ., Cent. for Wetland Resources, LSU-SG 74-01, 59 pp.

Kuzenski, Sally; Becker, Ronald. 1974. Crawfish: from bayou to big time. Texas A&M Univ., Sea Grant Coll. Prog, Sea Grant 70's 4(8):1-2.

MISCELLANEOUS FISH AND OTHER WORKS

Ables, Alan. 1972. The white amur . . . a star. Amer. Fish Farmer World Aquacult. News 3(6):4-5.

Allen, George; Conversano, Guy; Colwell, Bryan. 1972. A pilot fish pond system for the utilization sewage effluents. Calif. State Univ., Humboldt, Mar. Advisory Ext. Serv., Sea Grant Prog., CSUH-SG-3, 25 pp.

Anderson, Lee G. 1973. An economist looks at mariculture (the assessment of human needs and some problems of applying technology: the mariculture case). Mar. Tech. Society 7(3):9-15.

Anonymous. 1972. Annual report, 1972. Primary Prod. Dep., Ministry of Natl. Dev., Singapore, 38 pp.

Anonymous. 1973. Aquaculture in Canada, the opportunities and the risks. Fishermen and Ocean Science, 59(4):18-21.

Anonymous. 1973. Fish farming internationl. Vol. 1, Fish. News (Books) Ltd., Surrey, UK, 152 pp.

Anonymous. 1969. Fish farming today, a rapidly expanding multi-million dollar business. Amer. Fish Farmer 1(1):11.

Anonymous. 1973. Fish from farming. Fish Farm. Internatl. 1(1):14-15.

Anonymous. 1973. International development in fish production. O.E.C.D., Paris, 179 pp.

Anonymous. 1973. NOAA aquaculture survey, 1972, summary report to participants. (Also called the Mardella report.) Burlingame, Calif., Mardela Corp., a Sea Grant project report, 66 pp.

Anonymous. 1972. Proceedings, fourth national Sea Grant conference, Madison, Wis., Oct. 12 to 13, 1972. Univ. Wis., Sea Grant Communications Office, Sea Grant publ. WIS-SG-72-112, 248 pp.

Anonymous. 1972. Proceedings, Kauai aquaculture conference, Lihue, Kauai, June 25, 1972. Office of Econ. Dev., County of Kauai, and Dep. of Planning and Econ. Dev., State of Hawaii, 39 pp.

Anonymous. 1973. Report of the Pacific Island mariculture conference. Univ. Hawaii, Hawaii Inst. Mar. Bio., Coconut Island, Kanehoe, Hawaii, Feb. 6-8, 1973, 21 pp.

Anonymous. 1973. Rural recreation enterprises for profit (an aid to rural area development). U.S. GPO for U.S. Dep. Agri., Agri. Info. Bull. 277, 44 pp.

Anonymous. 1973. Sea fish farming progress. Fish Ind. Review. 3(2):2-6.

Anonymous. 1972. Summary of proceedings for the aquaculture conference, Honolulu, Hawaii, Nov. 9 to 10, 1972. Univ. Hawaii, cent. for Engineering Res., 60 pp.

Bardach, John E., Jr.; Ryther, John H.; McLarney, William O. 1972. Aquaculture: The Farming and husbandry of fresh-water and marine organisms. New York: John Wiley and Sons, Inc., 868 pp.

Bardach, John. 1972. Some remarks on aquaculture. Proceedings, Fourth Natl. Sea Grant Conf., Madison, Wis., Oct. 12 to 13, 1972, Univ. Wis., Sea Grant Communications Office, Sea Grant publ. WIS-SG-72-112, pp. 83-88.

Bardach, John E.; Ryther, John H. 1968. The status and potential of aquaculture, parts 1, 2 and 3. Clearinghouse for Federal Sci. Technical Info., 531 pp.

Blumberg, Robert C. 1971. Massachusetts solons study proposed aquaculture law. Amer. Fish Farmer 2(9):5, 22.

Boozer, David. 1973. Tropical fish farming. Amer. Fish Farmer, 4(8):4-5.

Botsford, L.W., Rauch, H.E., and Shleser, R. 1974. Applications of optimization theory to the economics of aquaculture. Unpublished manuscript, Bodega Marine Laboratory, University of California, Davis.

Brown, Evan E. 1973. Mariculture and aquaculture. Food Technology, 27(12):60-66.

Dassow, J.A.; Steinberg, M.A. 1973. The technological basis for development aquaculture to produce low-cost food fish. Mar. Fish Review 35(11):6-13. (Reprint 1015.)

Davidson, Jack R. 1972. Economics of aquaculture development. Proceedings, Fourth Natl. Sea Grant Conf., Madison, Wis., Oct. 12 to 13, 1972, Univ. Wis. Sea Grant Communications Office, Sea Grant publ. WIS-SG-72-112., pp. 75-83.

Dobson, W.D. 1972. Aquaculture: economic feasibility in the Great Lakes area. Proceedings, Fourth Annual Natl. Sea Grant Conf., Univ. Wis., Oct. 12 to 13, 1972. Univ. Wis., Sea Grant Communications Office, Sea Grant publ. WIS-SG-72-112, pp. 89-98.

Fedyaev, V.E. 1971. Method of planning, recording and calculating the cost of pond fish (O metodlike planirovaniya, ucheta i kal'kulirovaniya sebestoimosti prodovoi ryby). Rybnoe Khozyaistvo, no. 12, pp. 74-77. (Copy of translation available on loan from NMFS.)

Fijan, N. 1971. Present status and prospects of fish culture in Yugoslavia. Proceeding, Symposium on New Ways of Freshwater Fishery Intensification, Ceske Budejovice, (Zech.) September 22 to 24, 1971. Vodnany, Czech., Fish. Res. Inst., pp. 17-21.

Gates, John M.; Matthiessen, George C. 1971. An economic perspective. Thomas A. Gaucher, editor, aquaculture: a New England perspective. Univ. R.I., New Eng. Mar. Resources Info. Prog., (NEMRIP), pp. 22-50.

Gates, J.M. 1972. Appraising the feasibility of fish culture. Economic aspects of fish production, International Symposium on Fish. Econ., Paris, Nov. 29 to Dec. 3, 1971, O.E.C.D., pp. 327-348.

Gates, J.M. 1971. Aquaculture in less developed countries: some economic considerations. Preprints of 7th Annual Conf. of Mar. Tech. Society, Wash., D.C., Aug. 16 to 17, 1971, Mar. Tech. Society, pp. 579-583.

Gaucher, Thomas A., editor. 1971. Aquaculture: a New England perspective. Univ. R.I., Narragansett, R.I., New Eng. Mar. Resources Info. Prog., (NEMRIP), 119 pp.

Goodwin, Harold L. 1973. Aquaculture in perspective. Paper presented at the Sea Grant Assn. meeting, Oct., 1973, U.S. Dep. Commerce, Natl. Oceanic Atmospheric Administration, Natl. Sea Grant Prog., 17 pp.

Goodwin, Harold L. 1973. The aquaculture state of the art: comments on problems and progress. Presented at FISH EXPO, New Orleans, La., Nov. 26, 1973, unpubl., 15 pp.

Griffin, Charles. 1970. Loans for fish farmers and the types available. Amer. Fish Farmer World Aquacult. News 1(5):18-20.

Hammack, Gloria M. 1971. Bibliography of aquaculture. Coastal cent. for Mar. Dev. Serv., Wilmington, N.C., publ. 71-4, 245 pp.

Hudson, Stanton. 1974. Minnow farming, an American enterprise, then-now-and the future. Catfish Farmer World Aquacult. News 6(1):31, 32, 37 and 38.

Idyll, C.P. 1973. Marine aquaculture: problems and prospects. Technical Conf. on Fish. Management and Dev., Vancouver, Canada, Feb. 12 to 13, 1973, FI:FMD/73/S-37, Food and Agri. Org. of the U.N., Rome, 8 pp.

Insull, A.D.; Richardson, I.D. 1971. Economic assessment of marine fish development projects. Economic aspects of fish production, Internatl. Symposium on Fish. Econ., Paris, Nov. 29 to Dec. 3, 1971, O.E.C.D., pp. 319-326.

Iversen, E.S. 1968. Farming the edge of the sea. London: Fishing News (Books) Ltd., 307 pp.

Jones, A. 1972. Marine fish farming: an examination of the factors to be considered in the choice of species. Min. Agri. Fish., Food, Lab. UK, Leafl. (new ser.) 24, 16 pp.

Jones, Walter. 1970. Commercial fish farming: how to get started. Amer. Fish Farm. World Aquacult. News 1(2):5-8. Reprinted in 1972, 4(1):10-13, 19.

Joyner, Timothy. 1973. How can aquaculture help to preserve and build Alaska's fisheries? A paper presented to the Workshop on Aquaculture Potential in Alaska, Sitka, April 10-13, 1973. Nat. Mar. Fish. Serv., NW Fish. Cent., 6 pp.

Joyner, Timothy; Richards, Jack A.; Tanonaka, George A. 1971. Improving productivity of Washington's water resources through aquaculture. Natl. Mar. Fish. Serv., NW Fish. Cent., unpubl. manuscr., 27 pp.

Joyner, Timothy; Safsten, C. Gunnar. 1971. Prospects for sea farming in Pacific Northwest. Mar. Fish. Review 33(9):22-26.

Kensler, Craig B. 1970. The potential of lobster culture. Amer. Fish Farmer World Aquacult. News (11):8-12, 27.

Krupauer, V. 1971. Czechoslovak fishery and piscicultural research. Proceedings, Symposium on New Ways of Freshwater Intensification, Ceske Budejovice, Sept. 22 to 24, 1971. Vodnany, Czechoslovakia, Fish. Res. Inst., pp. 110-124.

Lewis, William M.; Heidinger, Roy. 1971. Aquaculture potential of hybrid sunfish. Amer. Fish Farmer World Aquacult. News 2(5):14-16.

Ling, S.W. A review of the Status and Problems of Coastal Aquaculture in the Indo-Pacific Region. In T.V.R. Pillay ed. Coastal Aquaculture—Proceedings of the IPFC Symposium on Coastal Aquaculture 1970. London: Fishing News Books Ltd., 1973.

Martekhov, P.F. 1966. On increasing the fish productivity of large lakes (Opyt polucheniya vysokoi ryboproduktivnosti na krupnykh ozerakh). Rybnoe Khozyaistvo, no. 7, p. 23. (Copy of translations available on loan from NMFS.)

Matthiessen, George; Gates, John. 1973. Aquaculture in New England. Univ. of R.I., New England Mar. Resources Info. Program (NEMRIP), 77 pp.

McNeil, William J., editor. 1970. Marine aquaculture. Selected papers from Conf. on Mar. Aquacult., Ore. State Univ., Mar. Sci. Cent., Newport, Ore., May 23 to 24, 1968. Ore. State Univ. Press, 172 pp.

McNeil, William J. 1971. Ocean food resources and future of mariculture. Proceedings, Academic Seminar, Jan. 26, 1971, Ore. State Univ., 11 pp.

Nash, Colin. 1972. Marine fish farming development. Proceedings, Kauai Aquacult. Conf., Lihue, Kauai, June 25, 1972, pp. 22-28.

Neal, R.A. 1973. Alternatives in aquacultural development: consideration of extensive versus intensive methods. Tech. Conf. on Fish. Management and Dev., Vancouver, Canada, Feb. 13-13, 1973, FI:FMD/73/C-2, Food and Agri. Org. of the U.N., Rome, 8 pp.

Odum, William E. 1973. The potential of pollutants to adversely affect aquaculture. Proceedings, 25th Annual Session, Gulf and Caribbean Fish. Inst., Miami, Fla., Nov. 1972. Univ. Miami, Rosenstiel School of Mar. Atmospheric Sci., pp. 163-174.

Palmer, H.V.R., Jr. 1974. Biologist finds sewage, aquaculture compatible. Natl. Fisherman, 55(2):12-B.

Pawlowski, R.R. 1972. Farming crocodiles down under (crocodile farming in Australia). Amer. Fish Farmer World Aquacult. News 3(10):12-15.

Philippines Fisheries Commission, Development Bank of the Philippines Fisheries Industry Development (A loan proposal to the World Bank), April 1972.

Pillay, T.V.R. 1973. The role of aquaculture in fishery development and management. J. Fish. Res. Board Can. 30(12):2202-2217.

Prewitt, Roy. 1971. Rambling along. Amer. Fish Farmer World Aquacult. News 2(12):17-18.

Prewitt, Roy. 1972. Rambling along. Amer. Fish Farmer World Aquacult. News 3(7):16-19.

Priddy, John M.; Culley, Dudley D. 1972. Frog culture industry. Amer. Fish Farmer World Aquacult. News 3(9):4-7.

Pritchard, G.I. 1973. Constraints to aquacultural development. Canadian Fisherman and Ocean Sciences 59(4):22-27.

Pritchard, G.I. 1973. With the decline of natural fish stocks other sources must be used. Western Fisheries 86(4):38-43.

Pryor, Taylor A. 1971. Hawaii and aquaculture: the blue revolution. Hawaii Dep. of Planning and Econ. Devel., 13 pp.

Purvis, George M. 1972. Fresh water fisheries begun in the Azores. Amer. Fish Farm. Aquacult. News 3(4):4-6.

Richards, Jack; Tanonaka, George. 1971. Mariculture in the United States. Natl. Mar. Fish. Serv., NW Fish. Cent., unpubl. manuscri., 9 pp.

Rouzaud, Pierre. 1973. General situation of aquaculture in France. Brackishwater Aquacult. in the Mediterranean Region, FAO, Stud. Rev. GFCM, (52):25-33.

Rutka, Justin. 1969. Evolution of public policies affecting exclusive use of coastal zone fishery resources: a comparison of public policies of Japan and the United States with implications on the status and potential of aquaculture in Hawaii. Univ. Hawaii, Dep. Agri. Econ., unpubl. paper, 27 pp.

San Feliu, J.M. 1973. Present state of aquaculture in the Mediterranean and South Atlantic coasts of Spain. Brackishwater aquacult. in the Mediterranean Region, FAO, Stud. Rev., GFCM, (52):1-24.

Sarig, S. 1972. Fisheries and fish culture in Israel in 1971. Bamidgeh 24(3):55-75.

Scott, Anthony. 1970. Economic obstacles to marine development. Mar. Aquacult. (selected papers from) Conf. on Mar. Aquacult., Ore. State Univ., Mar. Sci. Cent., Ore. State Univ. Press, pp. 153-167.

Shang, Yung C. 1973. Comparison of the economic potential of aquaculture, land animal husbandry and ocean fisheries: the case of Taiwan. Aquacult. 2(4):187-195.

Shang, Yung Cheng; Iversen, Robert T.B. 1971. The production of threadfin shad as live bait for Hawaii's skipjack tuna fishery: an economic feasibility study. Univ. Hawaii, Econ. Res. Cent., 42 pp.

Shpet, G.I. 1972. Comparative efficiency of fish culture and other agricultural activities per unit of area used. Hydrobiological J. 8(3):46-51.

Tal, S. 1973. New prospects in fishculture in Israel. Bamidgeh 25(3):67-71.

Trimble, Gordon M. 1972. Legal and administrative aspects of an aquaculture policy for Hawaii: an assessment. Hawaii, Dep. Planning and Econ. Devel., Cent. for Sci. Policy and Tech. Assessment, and Western Interstate Comm. for Higher Educ., Resource Internship Prog., 61 pp.

United Nations, Food and Agricultural Organization, Indo-Pacific Fisheries Council. 1970. *Report of the first meeting of the IPFC working party on economics of fish culture.* Bangkok.

United Nations, Food and Agricultural Organization, Regional Office in Bangkok. 1973. Preliminary analysis of questionnaires, IPFC working party on economics of aquaculture.

U.S. Dept. Interior, Bur. Sport Fish. Wildlife. 1970. Report to the fish farmers (the status of warmwater fish farming and progress in fish farming research). U.S. Dept. Interior, Bur. Sport Fish. Wildlife, Resource publ. 83, 124 pp.

Webber, Harold H. 1973. Risks to the aquaculture enterprise. Aquacult. 2(2):157-172.

Welsh, James P. 1974. Mariculture of the crab Cancer Magister (DANA) utilizing fish and crustacean wastes as food. Humboldt State Univ., Sea Grant Project, HSU-SG-4, 76 pp.

White, James T. 1970. Minnows—by the million. Amer. Fish Farmer World Aquacult. News, August, 1970, pp. 8-11.

Winget, Rodner R.; Maurer, Don; Anderson, Leon. 1973. The feasibility of closed system mariculture: preliminary experiments with crab molting. Proceedings, Natl. Shellfish Assn. 63 (June), reprint, Univ. Del.

Index

About the Authors

Frederick W. Bell has an international reputation as a fisheries economist. His numerous publications also include works in the fields of resource and environmental economics, regional economics, banking and finance, and industrial organization. His articles have appeared in the *Journal of Political Economy, American Economic Review, Southern Economic Journal, Review of Economics and Statistics, Journal of Finance,* and *the Quarterly Journal of Economics.* He is an Associate Editor of the *Journal of Regional Science.* Professor Bell is a grant recipient from the U.S. State Department, U.S. Nuclear Regulatory Agency, National Commission on Water Quality, and the U.S. Department of Commerce. He has served as a consultant to the U.S. Department of Transportation, Fisheries Development Limited, and Synergy, Inc.

E. Ray Canterbery is well-known for his publications in the field of international economics. In addition to *Foreign Exchange, Capital Flows, and Monetary Policy,* Studies in International Finance, Princeton University Press, 1965, he also is the author of the critically-acclaimed *Economics on a New Frontier,* Wadsworth, 1968, and *The Making of Economics,* Wadsworth, 1976. He also has authored numerous articles in such journals as the *American Economic Review, Journal of Political Economy, Quarterly Journal of Economics, American Journal of Agricultural Economics, Canadian Economic Journal* and the *Southern Economic Journal.* Professor Canterbery has served as a consultant to the U.S. State Department and has been the recipient of grants from that agency as well as the Rockefeller Foundation, the Nuclear Regulatory Commission, and the National Commission on Water Quality.